SCULPTUR
CASTING

Other Books by Dona Z. Meilach
in Crown's Arts and Crafts Series

CONTEMPORARY BATIK AND TIE-DYE
MACRAMÉ; CREATIVE DESIGN IN KNOTTING
MACRAMÉ ACCESSORIES
PAPIER-MÂCHÉ ARTISTRY
CONTEMPORARY STONE SCULPTURE
CONTEMPORARY ART WITH WOOD
DIRECT METAL SCULPTURE
with Donald Seiden
SCULPTURE CASTING
with Dennis Kowal

Also:

CONTEMPORARY LEATHER: ART ACCESSORIES
CREATING ART FROM ANYTHING
CREATING WITH PLASTER
MAKING CONTEMPORARY RUGS AND WALL HANGINGS
PRINTMAKING
PAPERCRAFT
ACCENT ON CRAFTS
CREATIVE STITCHERY
with Lee Erlin Snow
COLLAGE AND FOUND ART
with Elvie Ten Hoor

SCULPTURE CASTING

Mold Techniques and Materials
Metals · Plastics · Concrete

by

DENNIS KOWAL and DONA Z. MEILACH

CROWN PUBLISHERS, INC., NEW YORK

Everyone will tell you I am not a musician. (cf. O. Séré: Musiciens français d'aujourd'hui, *p. 138). That is true. From the beginning of my career I classed myself among the "phonometrographers." My work is nothing but pure "phonometry." Take, for example, the* Fils des Étoiles, *or the* Morceaux en forme de Poire, En Habit de Cheval, *or the* Sarabandes, *and it will be seen at once that in the creation of these works musical ideas played no part at all. They are purely scientific. And as a matter of fact it gives me more pleasure to measure a sound than to hear it. With a phonometer in my hand I work happily and surely. What is there that I haven't weighed or measured? All Beethoven, all Verdi, etc. It's most interesting. The first time I used a phonoscope I examined a B flat of average dimensions. I can assure you I never in my life saw anything quite so repulsive. I had to call my servant to come and look at it.*

On my phono-weighing-machine an ordinary F sharp, of a very common species, registered 93 kilograms. It came out of a very fat tenor whom I also weighed.

Do you know how to clean sounds? It is rather a dirty process . . .

—*from* Erik Satie *by Rollo H. Myers*

Inquiries should be addressed to Crown Publishers, Inc.,
419 Park Avenue South, New York, N.Y. 10016.

Library of Congress Catalog Card Number: 72-84319
ISBN: 0-517-500590

Printed in the United States of America
Published simultaneously in Canada by
General Publishing Company Limited

Acknowledgments

THE monumental task of gathering material for *Sculpture Casting* was made possible by the cooperation of literally hundreds of people in various parts of the world. Our thanks go to all the people in industry or with museums and galleries who corresponded with us regarding photos, supplies, and techniques. Where possible, credit accompanies the photos.

Our sincere admiration and appreciation is extended to the many photographers whose expertise contributed to the total visual quality of this book.

Above all, we are deeply indebted to the artists who submitted photos of their work or who permitted us to photograph their sculptures. We especially want to thank those who chronicled lengthy procedures over a period of several months so that we would have a photographic record of the steps and the problems that had to be solved; and to those who graciously volunteered to do demonstrations for us to photograph. These include: Paul Aschenbach, Stephen Daly, Peter Fagan, Tom Goldenberg, Phillip Grausman, Bradford Graves, Dimitri Hadzi, Leo Kornbrust, Roger Kotoske, Jacques Schnier, Albert Vrana, Thomas Walsh, and Paul Zakoian. We also extend our appreciation to George Greenamyer and his students at the Hinckley-Haystack School of Crafts.

We are grateful to Lebbeus Woods of Champaign, Illinois, for his excellent interpretation of the technical drawings.

We especially want to thank our spouses and children for their patience and understanding when we spent hours, days, and months organizing and compiling the photos and text into a meaningful book.

Finally, we should thank one another for putting up with the vicissitudes of our individual natures and moods for the duration of this project.

DENNIS KOWAL, Boston, Mass.
DONA MEILACH, Palos Heights, Ill.

Note: All photographs by Dennis Kowal unless otherwise noted.

Foreword

Sculpture Casting explains in detail traditional and modern systems for mold making and casting. Working methods by various artists will help the reader appreciate how others have solved many of the problems that arise in sculpture casting.

All demonstrations are aimed to explain materials and methods and how they are used. Becoming familiar with many approaches will enable the sculptor to be more versatile and flexible. Artists, faced with their own problems of casting, are encouraged to collaborate with manufacturers, distributors, chemists, engineers, and other artists to solve some of the technical questions. Often technical libraries and industrial sources must be consulted to find answers.

This book is intended for the student and the professional sculptor. The scope of the book includes the basic concepts of various mold marks and casting techniques and materials. All can be altered, reapplied, and expanded into other areas; they provide a solid base for further development. Many examples of finished works give ample testimony to the validity of the casting method for contemporary sculpture.

DENNIS KOWAL
DONA Z. MEILACH

SCULPTURE CASTING

Disco #1. Arnaldo Pomodoro. Polished bronze on a black metal base, 27-inch diameter.
Marlborough-Gerson Gallery, Inc., New York

1

Sculpture Casting

THE principle of casting is as old as nature. The fossilized replicas of shells, fish, and plants are the result of natural forces incasing an original with a material such as sand or mud which becomes a mold. The original, after millions of years, becomes fossilized and the mud turns to rock. By splitting the rock open the fossil is revealed, one side of the rock having a negative impression, the other a positive. A casting or reproduction can then be made by pouring another material into the negative half of the stone.

Casting today is based on the same principles. It begins with an original form, usually of a relatively impermanent material, such as foam, wax, clay, or plaster. The mold is made from plaster, plastic, rubber, gelatin, or other suitable materials that will copy the surface and form of the original. The original is removed; the mold is prepared and the casting material is poured into the mold. The casting material can be clay, wax, plastic, metal, or concrete. Bronze, traditionally, is the most popular metal for casting, but iron, copper, gold, lead, aluminum, pewter, and combinations of metals are also used.

Until the 20th century casting was essentially done using metals and plaster, re-creating originals by the techniques man had used for centuries. The oldest and most well known method of casting metals is called "cire perdue" or "lost wax." It was practiced in ancient Egypt and Greece and by the Indus Valley cultures.

In the last few decades plastics have moved into the casting sculptor's studio as a material for making molds and for creating finished castings. The result of the sculptor's innovative use of materials initiated by and for industry has drastically changed the practice and direction of sculpture casting with exciting results. Much of the interest and activity has been fostered by sculpture conferences at the University of Kansas, Lawrence, Kansas, under the direction of Eldon Teft.

Until recently the casting processes were costly and made to appear complicated. The sculpture conferences, which originally dealt with casting metals, revealed that the sculptor could cope easily with foundry procedures on a smaller scale than did industry and, therefore, be able to carry his work through the entire creative process himself. Instead of fashioning his work in wax, clay, or plaster and then turning it over to foundrymen who did not necessarily relate to the esthetics visualized by the sculptor, the sculptor could follow the creative and technical aspects through to the final finish himself.

The role of the artist in relation to his

PAIR OF HORSES. China. Shang dynasty, pre-An-yang period, 1523–1300 B.C. Bronze, 4½ inches high, 9½ inches long.

The Cleveland Museum of Art, Leonard C. Hanna, Jr., Collection

work has developed so that many are now involved with the materials and processes from beginning to end. It became apparent that the artist could develop his own foundry for castings that were not beyond certain sizes—perhaps twenty feet tall. A few artists set up their own foundries, first small, then larger, and discovered great advantages in doing the work themselves.

Today, artists and art schools are building foundries in garages, studios, or outdoors in relatively small spaces. Equipment costs are not prohibitive. Setups for installation of simple foundries are illustrated in this book.

In addition, more available foundries, some operated by sculptors, are permitting the artist to work closely with the development of his piece. Even if the artist chooses to have the foundry invest and pour the casting, he, himself, often prefers to do the final chasing and finishing.

The sculpture conferences, by combining the experience and talents of many artists, soon moved from metal casting to concrete, plastics, and, most recently, glass. They involved representatives of industry who explained the application of newer mold materials, among them RTV and the cold molding compounds, and of casting materials such as polyester resins and epoxies. They showed how they were used industrially and encouraged sculptors to experiment with the materials. The result was a variety of techniques and materials for casting that could be accomplished without complicated foundry procedures. It eventually led to new statements in new media. The trend, though still in its early stages, can be readily observed in gallery and museum shows throughout the world.

Ceramic shell casting is another technique borrowed from industry. While the molds are more complicated than those used for lost-wax casting, the procedure is recommended for certain types of sculpture.

Sand casting, the most basic industrial method, is used for casting all metals and at times plaster and concrete.

Problems that artists have with molds

and pouring materials have sent them searching for solutions often found in the laboratories of chemists and the experience of engineers. Fortunately, industry is extremely receptive to the artist's problems, and it is from the collaborative efforts between artists and manufacturers that some of the newer techniques and materials have developed.

Because so many of the materials demonstrated and explained throughout the book are not normally stocked by regular artist supply sources, the reader must often search out his supplies from many different sources. A selective, but only partial, list of suppliers may be found in the Appendix. In addition, the classified advertisements in magazines, newspapers, and telephone directories are the best sources for local information. Almost all types of plastic suppliers can be found under Plastics and related entries; the same for metals, foundries, concrete, and tool outlets. In the local library, refer to *The Thomas Register* for additional suppliers around the country. And always ask questions of manufacturers and distributors.

Sculpture Casting is mainly intended to provide the student and professional with the technical information he needs for proceeding with all types of mold making and casting materials. An acquaintance with man's involvement with casting techniques and the results that have been achieved over thousands of years attest to the validity of casting as a method for making sculpture.

The refinement of ancient bronze castings is a staggering testimonial to the sophisticated methods used. The traditional rigid mold method used today has been used for thousands of years. Evidence is seen in the sculpture, pots, jewelry, weapons, and other useful items created prior to 3500 B.C.

WINE VESSEL. China. Late Shang or early Chou dynasty, 12th to 10th century B.C.

Art Institute of Chicago

Of all metals, bronze has been, and still is, the most widely used metal for casting sculpture. It is an excellent material for outdoor use because it is structurally strong and highly resistant to corrosion. Clean bronze is a shiny golden color; the ancient Greeks preferred the shine of the metal. Old bronzes that have a black, green, and brown color are exhibiting a patina that is the result of corrosive effects of weathering or of having been immersed in sea water or buried in the ground for centuries. Today's artists often purposely apply a patina to color bronze, using a variety of acids to achieve the desired effect.

Bronze is most often cast by the lost-wax method; sometimes wood and other organic materials are substituted for wax. It is amazing that the Chinese of the early Shang dynasty (about 1523-1028 B.C.) developed the lost-wax process to a highly specialized, technical level. The early Greek bronze sculptures were also created by this method and the Romans followed suit. Few western bronzes remain from earlier civilizations because bronze can be melted, which the Romans did, for example, when they melted sculpture down for cannonballs. Phoenician traders drove a thriving business in scrap bronze. Many of the important works that had survived from the ancient world were melted in the holocaust that followed the capture of Constantinople by the Crusaders in 1204, when statues were piled in churches waiting to be hauled to the foundries. A few survivors remain, among them, the famous four horses atop the entry of St. Mark's Cathedral in Venice.

During the Renaissance, and later, sculptors did insist on keeping the casting processes in their own studios. Ghiberti, for example, was his own bronze-founder, and Benvenuto Cellini, in his *Autobiography,* describes the casting of his large bronze sculpture *Perseus* in his own workshop.

The 16th-century bronze heads from Benin, Africa, are another chapter in the use of the lost-wax technique by a group of people who developed their own unique methods of lost-wax casting despite the lack of a sophisticated scientific knowledge of metals and furnaces. The discovery of the Benin bronzes in comparatively recent years was a remarkable event.

The history of sculpture can easily be followed through the bronzes that remain, and the dates and origins of those throughout the book should be observed. By the end of the 19th century, Auguste Rodin used smooth, reflecting, and patinated bronze surfaces, which many critics feel were the counterpart of the shimmering, light-breaking surfaces of his contemporaries, the Impressionists. With this innovation, he created a surge of new interest in bronze casting that is still being explored.

Moving into the 20th century, the story of bronze casting continues despite the development in the 1930s of direct metal sculpture—assemblages and constructions created by welding metal together rather than casting it. Actually, welding, brazing, and soldering soon were used along with casting methods to combine small portions of castings into monumental works.

By the 1950s and 1960s, casting moved out of the professional foundry into the studios and into the classrooms. As the Bronze Casting Conventions in Lawrence, Kansas, disseminated information, sculptors everywhere, hungrily seeking information, worked with the materials and methods that promised a multitude of new directions for casting. Information spread throughout the country and, soon, casting courses were being offered in the prominent art schools, then in the art departments of smaller schools. As so many plastics and flexible mold materials are being developed and improved, and the technology for creating

KARAIKKAL-AMMAIYAR, A SAINT OF SIVA. India. Chola period, 12th to 13th century. Bronze, 16¼ inches high.

Nelson Gallery of Art, Kansas City, Mo.

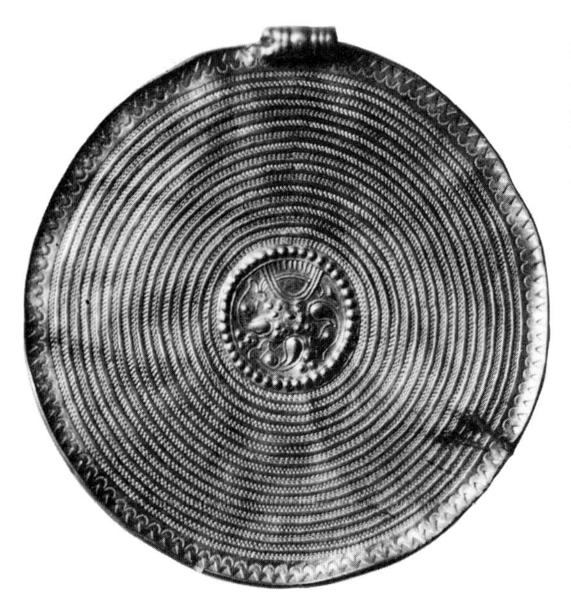

BRACTEATES. Kodings, Hemse, Gotland. 8th century. Decorated gold.
Museum of National Antiquities, Stockholm
Courtesy, Art Institute of Chicago

metal foundries becomes more easily available and less costly, casting as an important and continuing method for creating sculpture promises to become an increasing activity of sculptors everywhere.

This book is designed to help corroborate current enthusiasm for cast sculpture. It illustrates the technical know-how needed for working with a variety of materials. It offers historical examples to show the heritage of casting and contemporary ideas. In addition to offering technology for the serious student and professional, it is an important book for everyone interested in the various processes of the art.

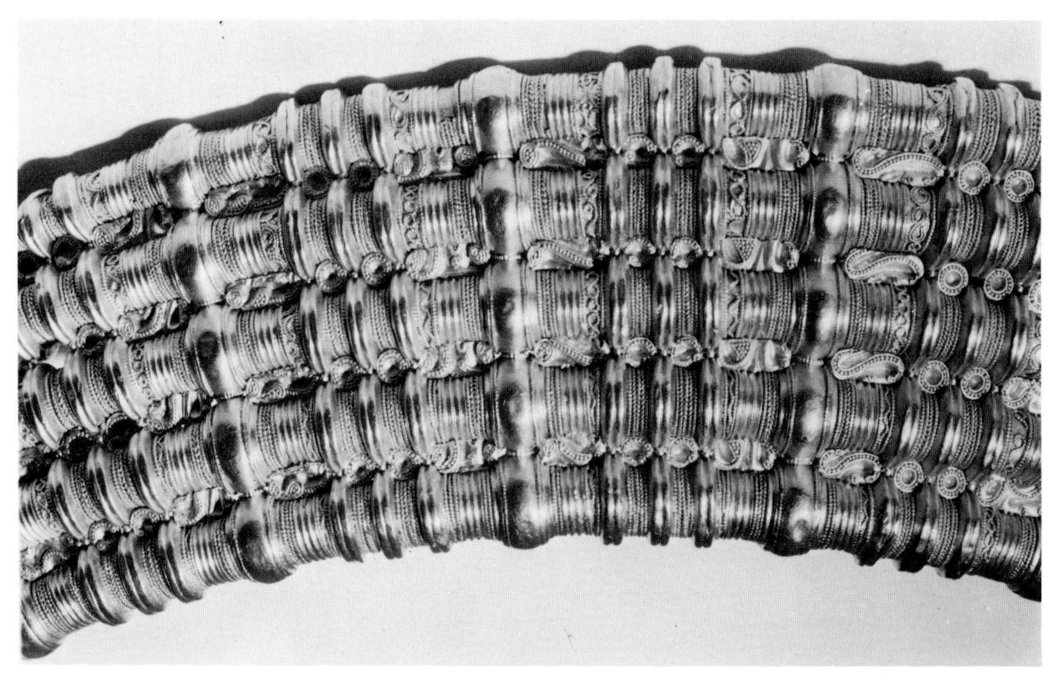

COLLAR. Färjestaden, Torslunda, Öland. Gold.
Museum of National Antiquities, Stockholm
Courtesy, Art Institute of Chicago

DAVID. Donatello. Ca. 1430. Bronze.
National Museum, Florence
Alinari-Art Reference Bureau

HERCULES AND ANTAEUS. Giovanni Bologna. Bronze.
Art Institute of Chicago

BRONZE HEAD. Africa (Benin). 16th century.

Field Museum of Natural History, Chicago

HEAD OF RENOIR. Aristide Maillol. Bronze, 16 inches high.

Art Institute of Chicago

BUST OF A NEGRESS. Charles Henri J. Cordier. Bronze.

Art Institute of Chicago

THE HAMMERMAN. Constantin Meunier. Bronze, 76½ inches high.
Art Institute of Chicago

LE GRAND CHEVAL. Raymond Duchamp-Villon. 1914. Bronze, 38 inches high, 48 inches wide, 21 inches deep.
Walker Art Center, Minneapolis

DYNAMO MOTHER. Gaston Lachaise. 1933. Bronze, 12 inches high, 18 inches wide, 7½ inches deep.

Museum of Modern Art, New York

FIGURE. Jacques Lipchitz. Bronze.
Los Angeles County Museum of Art

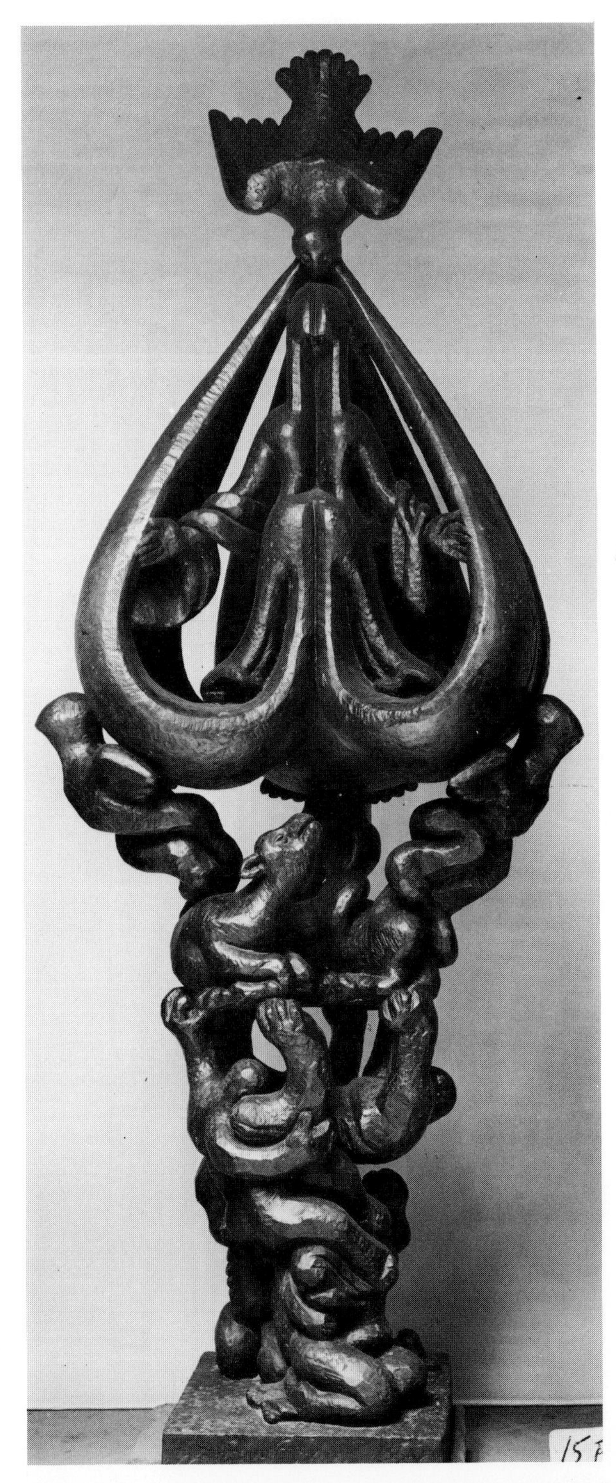

PEACE ON EARTH. Jacques Lipchitz, 1960.
Marlborough-Gerson Gallery, Inc., N.Y.

GRANDE TÊTE. Alberto Giacometti. Bronze.
Collection, Mr. and Mrs. Arnold Maremont, Winnetka, Ill. Courtesy, Art Institute of Chicago

DANCE OF SLEEP OR DEATH. Marianna Pineda. 1953. Bronze, 12 inches long.
Addison Gallery of American Art, Phillips Academy, Andover, Mass.

RECLINING MOTHER AND CHILD. Henry Moore. 1960–61. Bronze, 86½ inches long, 33 inches high, 54 inches wide.
Walker Art Center, Minneapolis

SOLAR BIRD. Joan Miró. 1966. Bronze.
Art Institute of Chicago

PASSAGE (detail). Harold Tovish. Bronze. *Courtesy, artist*

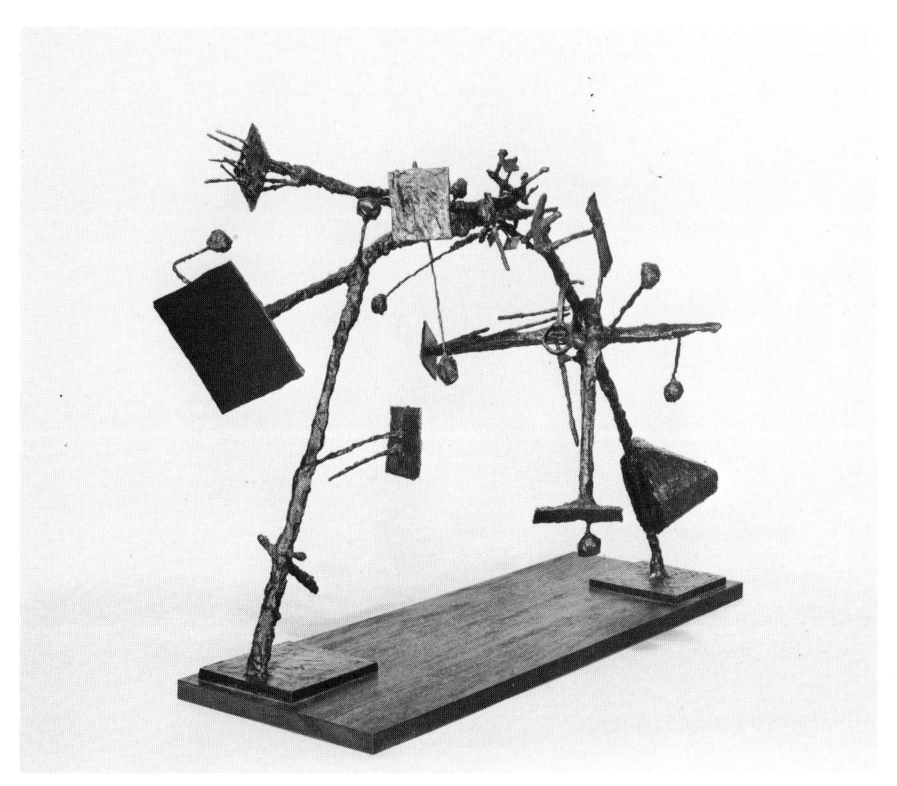

COUPLE IV. Matta. 1959–60. Bronze.
Art Institute of Chicago

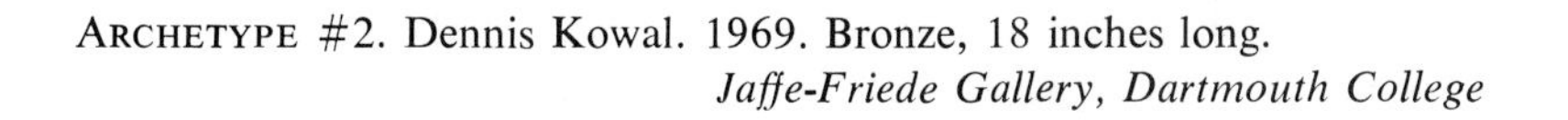

ARCHETYPE #2. Dennis Kowal. 1969. Bronze, 18 inches long.
Jaffe-Friede Gallery, Dartmouth College

MOTHER AND CHILD ON CHAISE LOUNGE. Rhoda Sherbell. 1970. Bronze.
Harbor Gallery, New York

Around and About. Clement Meadmore. Cast polyester, 5¼ inches high, 8 inches wide, 5 inches deep. Maquette for welded aluminum piece: 84 inches high, 132 inches wide, 87 inches deep.

Courtesy, artist

Sculpture. Marta Pan. Cast cement. *Courtesy, artist*

STANDING FEMALE FIGURE #2. Alba Corrado. 1972. Life size. Polyester resin, fiberglass, and velour.

Standing Gymnopédies. Dennis Kowal. 1972. Aluminum, 21¾ inches tall.

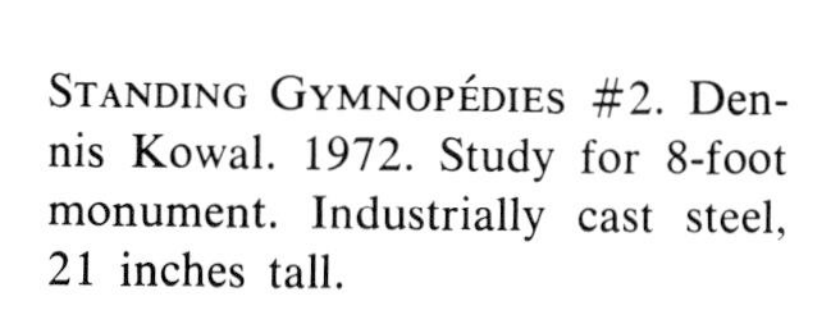

Standing Gymnopédies #2. Dennis Kowal. 1972. Study for 8-foot monument. Industrially cast steel, 21 inches tall.

2

Casting Polyester Resin with RTV Silicone Rubber Molds

CASTING sculpture has been greatly simplified by the development of plastic resins and fiberglass. Today, for the amateur or professional, polyester resins poured into flexible molds provide a standard casting technique that gives excellent results. The materials, their ease of use and availability, and the minimal space requirements make resin and flexible molds extremely appealing to the sculptor.

Polyester resin castings have been on the contemporary sculpture scene for several years and continue to increase in popularity. Polyester resin is a thermosetting plastic; it is resistant to chemicals, moisture, and corrosion, and is quite strong, lightweight, and less expensive than epoxy.

Popular flexible mold materials for casting polyester resins are vinyl, silicone rubber, polyvinyl chloride (PVC) plastisols, flexible epoxies, and natural latex. A silicone rubber mold in the Room Temperature Vulcanizing (RTV) family is often preferred, though expensive, because it releases castings easily, is heat resistant, and accurate. It is demonstrated in the accompanying photo series. PVC plastisols, vinyl, flexible epoxies, and natural latex are low in cost but have disadvantages. PVC plastisols break down under heat, vinyl shrinks excessively, and flexible epoxy releases castings poorly. Latex is attacked by the resin and latex molds take longer to make than the others.

The original pattern from which the RTV mold is made can be plaster, wood, plastic, clay, and other materials, as long as the surface is firm enough to be coated with RTV. RTV will reproduce even the most minute details and undercuts, so surface blemishes on the original should be removed or filled. Porous materials can be sealed by spraying with several coats of clear lacquer. Waxing and polishing the surface of the original will give a high luster to the plastic reproduction.

The polyester resin material is quite versatile and can be any color, transparent, translucent, or opaque. It may be formulated to produce almost any degree of flexibility but most are rigid. Degrees of rigidity are obtained by adding a styrene monomer as high as 30 percent, which is the state of most commercial products.

The density of polyester resin can also vary from approximately 55 to 80 pounds per cubic foot. By using a dense resin and adding a filler such as calcium carbonate, a marble-like casting can be made. Density can be reduced by adding fillers such as

UNTITLED. Roger Kotoske. 1970. Transparent colored polyester resin, 10 inches high, 20 inches wide. Finished casting made from an RTV silicone rubber mold.

glass microspheres, ceramic spheres, and vermiculite, which also strengthens the cast and changes its appearance.

When casting polyester resins, the exotherm, or heat generated during catalyzing, is relatively low and can be controlled by the resin producer to a degree in formulation. It can also be controlled by the caster by regulating the amount of catalyst used; the more catalyst, the more heat. Casting thick sections (over one inch thick) can generate excessive exotherm. Increasing the exotherm can ruin a mold or cause cracks and crazing in or on the casting. Reducing the exotherm will extend the gel time.

Castings are usually reinforced with fiberglass cloth, mat, or fibers; metal wire or rods; and wood.

The addition of fillers, lighteners, and reinforcements will also help check shrinkage during curing. Shrinkage runs between 2 and 2½ percent in standard polyester. The recently developed water-filled types of resins shrink after curing; this "post-cure shrinkage" results from the evaporation of the water filler. These new resins are being refined by the furniture industry for furniture production.

Other types of polyester resin are called promoted or non-promoted. The promoted type requires the addition of fillers and catalyst, and the gel or setting time is determined by temperature and/or the amount of catalyst, which is normally MEK peroxide. With the non-promoted type, add a promoter that fits your requirement. The promoter is the agent that can speed the reaction between various resin ingredients. Do not confuse it with the catalyst. You must consider that a promoter and catalyst may not work together properly. A typical promoter is benzoyl peroxide catalyst and diethyl aniline and cobalt promoter, or MEK peroxide catalyst and cobalt promoter.

All resins require certain precautions to assure a good casting. Do not use a silicone spray mold release if the casting is to be painted. Avoid air bubbles while mixing the components. Cast into dry molds—water is always to be avoided. Allow air bubbles to escape through various means such as working on a vibrating table. Preplan the filler and color to be used for compatability and amount, measure accurately, and mix well. Always know your materials and *read directions thoroughly* before proceeding. Always adhere to safety precautions suggested by the manufacturer.

In the accompanying demonstration, Roger Kotoske uses a Silastic D RTV Mold Making Rubber distributed by Perma-Flex Mold Company, Columbus, Ohio. Similar products are available from Dow, Corning, General Electric, and others.

PREPARING THE ORIGINAL MODEL

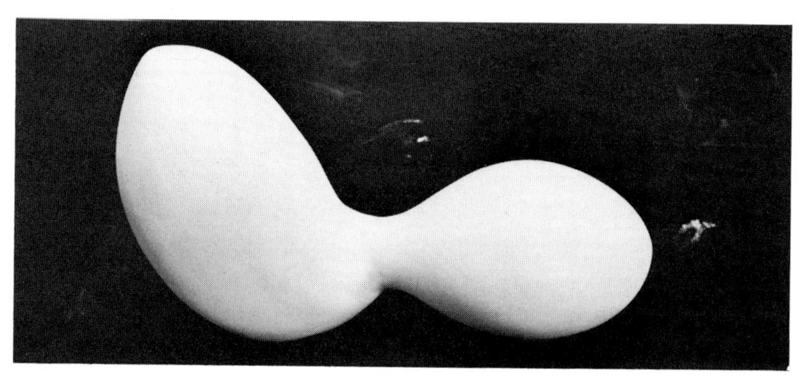

The original form to be cast is made of plaster. The mold will be RTV silicone rubber.

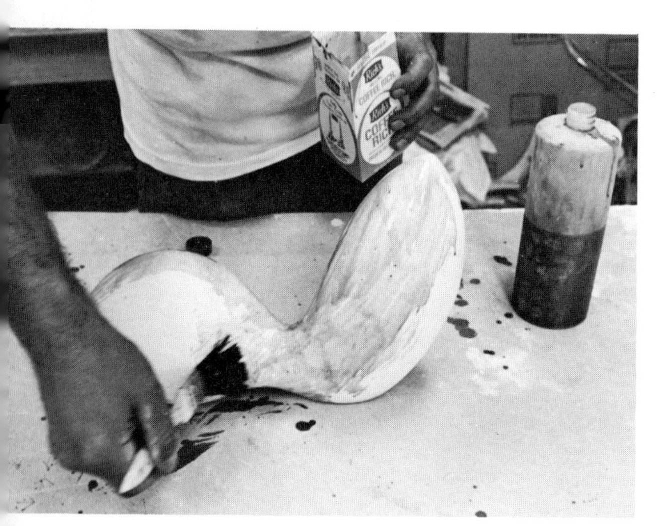

The finished plaster surface is coated with 8-ounce orange shellac cut with methanol thinner. Shellac is attacked by RTV but will be protected with one coat of PVA.

Any holes or imperfections should be patched with plasteline clay because RTV reproduces minute surface details.

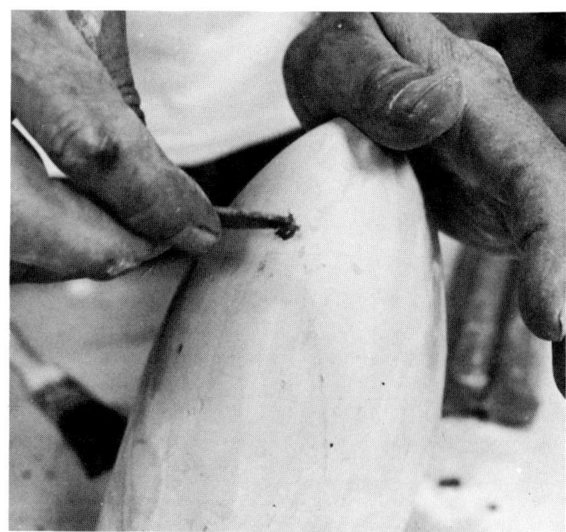

Find a center line and begin to divide the form in half with wet clay. Where multiple pieces of a mold are to be made, divide form and mark to indicate mold separations.

Continue to build up clay walls. If the piece will require extensive work after casting, great surface accuracy is not necessary other than seeing that mold and casting are true to shape. Register keys are then made. Slightly roughen the clay flange surface to help prevent "casting" material from leaking out of mold.

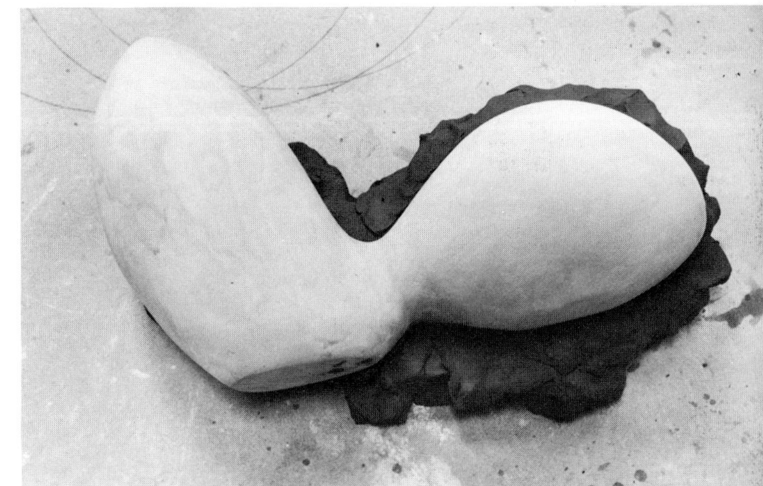

Coat plaster and wet clay with polyvinyl alcohol (PVA) using a soft brush, especially where a retouching coat is called for, as the first film is delicate. Wet clay requires a more delicately brushed coat of PVA than plaster and wood. When brushing, avoid puddles, use warm dry air to dry, and attain an even coat. If the original is made of wet clay it would be best to make a plaster cast of the original, then make the RTV mold from the plaster.

USING RTV

To prepare RTV, pour it into a container (cardboard will do), and measure accurately by weight to determine the amount of curing agent to be added. After weighing, the catalyst is added. Follow product directions. Mix thoroughly by hand using a spatula or a paint stick until only one color remains. A power mixer can be used. Always scrape the sides of the container clean several times while mixing.

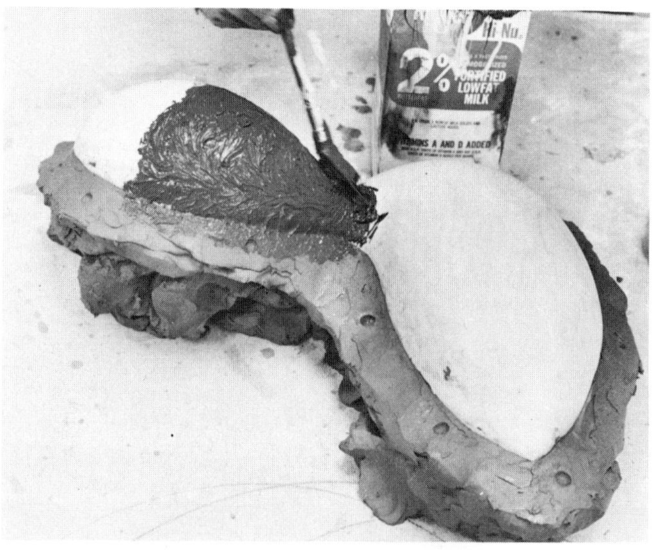

Apply mix with a brush appropriate to the size of the piece. Apply in a circular or swirling motion from the center out to help eliminate surface air bubbles. Brush until entire piece and parting line are coated about ½ inch thick. (Thickness is a matter of personal preference but is usually determined by the size and complexity of the piece and the support mold to be used.)

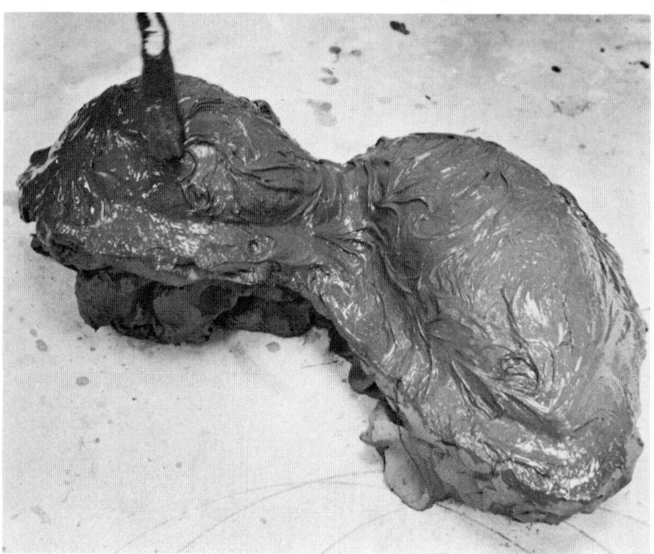

The flange thickness should be enough to resist tearing or flexing and provide accurate registering.

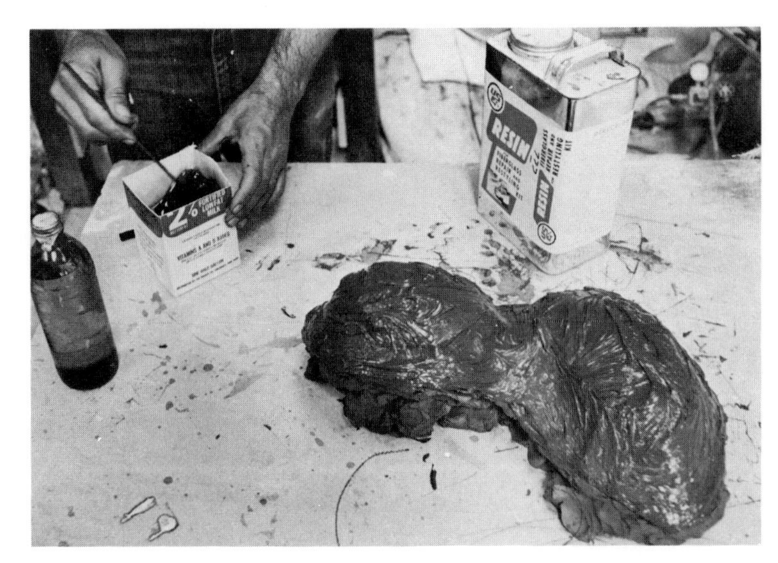

If air pockets remain on the finished mold, repair with silicone rubber from tube of bathtub sealer. After the initial coat is applied, the remaining material can be added with less concern, as the first layer captures the form and surface detail, the second generally adds strength.

MAKING THE SUPPORT MOLD

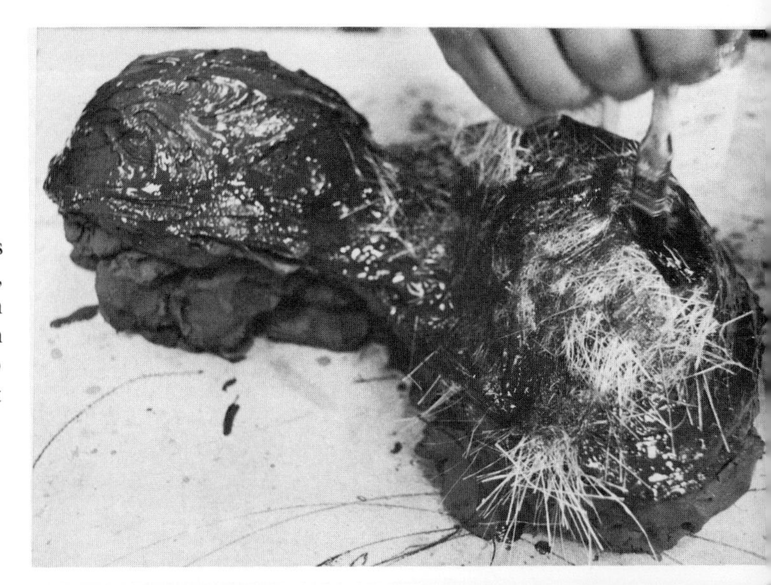

The first section of the support mold is made with polyester resin and fiberglass, mat or cloth, as reinforcement. Brush on one coat of resin and immediately lay on fiberglass. Three layers of resin and two of fiberglass will make a strong support mold. Mix resin for fast setup.

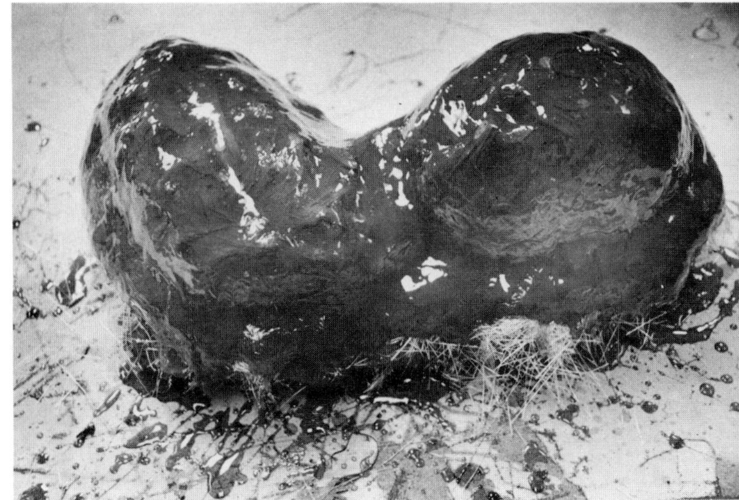

Polyester resin and fiberglass mat section completed, except for trimming excess mat along edges.

Remove support mold by gently prying it loose from the RTV. Trim edges and replace. Turn piece over to make second half of mold. Remove the clay divider. Clean the rubber flange and the other half of plaster form now to be covered.

Determine where pouring holes will be located. High points should be level and marked so when casting material is poured, air and heat can escape.

Plugs for pouring holes are tapered with the wider diameter on the form to facilitate removal of mold. Coat flange, plaster, and pouring holes with PVA.

To make the second half of the RTV mold, follow the same procedures as for first half.

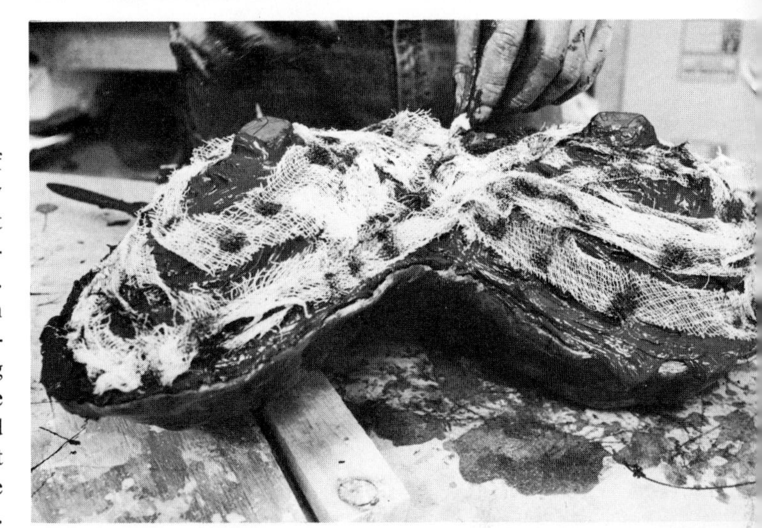

On the final coat of RTV add a layer of gauze to the surface, leaving it partially impregnated with the RTV. Areas that are not impregnated will lock to and become part of the polyester support mold. This prevents the upper mold section from collapsing during pouring and casting. Other ways of temporarily locking the flexible and support molds may be used: plugs, wire, or string in the mold that are tied to the outside of the support mold. The complexity of the piece to be cast will determine how the mold is made.

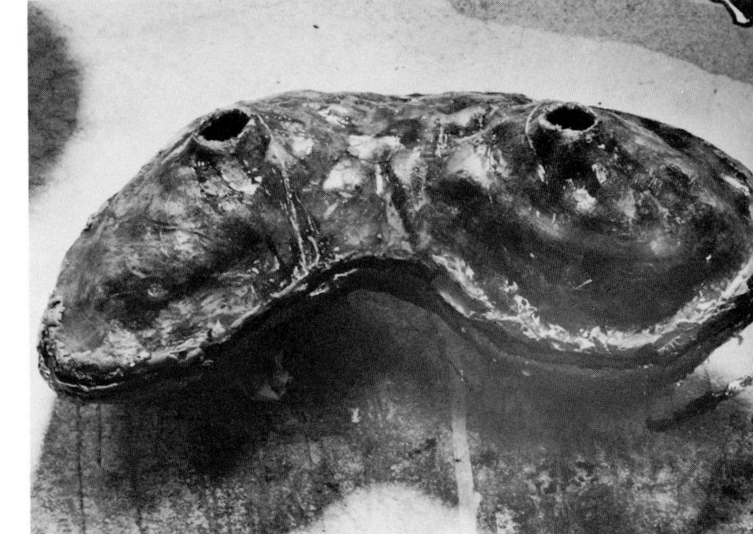

Finished support mold.

Open mold. Check register of both the flexible and the support mold. Clean and coat the inside of the flexible mold with a release agent such as wax, PVA, or silicone, depending on casting material to be used. Some materials do not require a release agent. Always check and recheck for dirt and other foreign particles inside the mold.

Left: The open, completed mold showing pouring holes in the top section. *Right:* Another completed mold with plaster support mold.

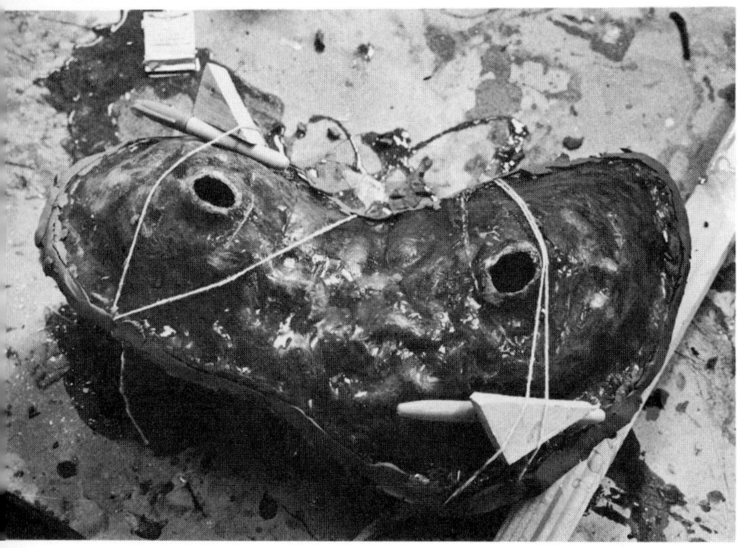

Tie mold together with string or rubber strips. Patch steams with clay. Level and stabilize piece for pouring. Be sure ties are taut.

POURING THE CASTING MATERIAL

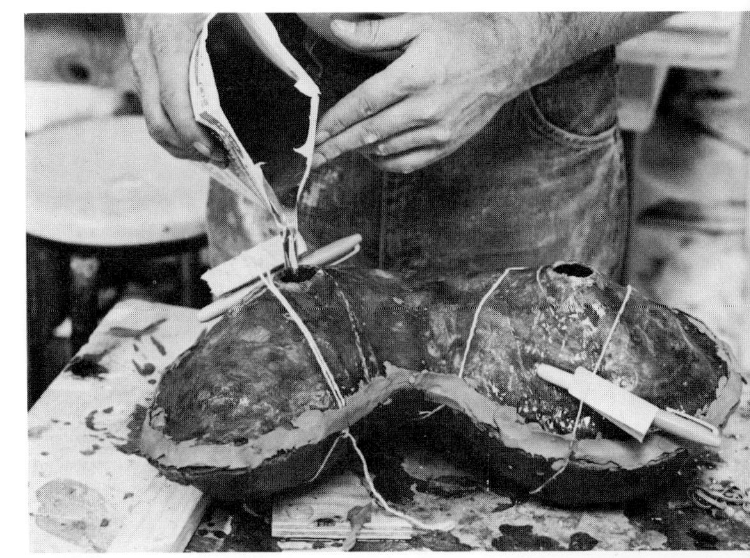

Pour polyester resin tinted with a liquid dye into mold. Use less catalyst for a slow cure to avoid heat, bubbles, and cracking. One-half inch to one-inch resin pours are normal, but some plastics can be cast solid. Always check and know the potential of your materials.

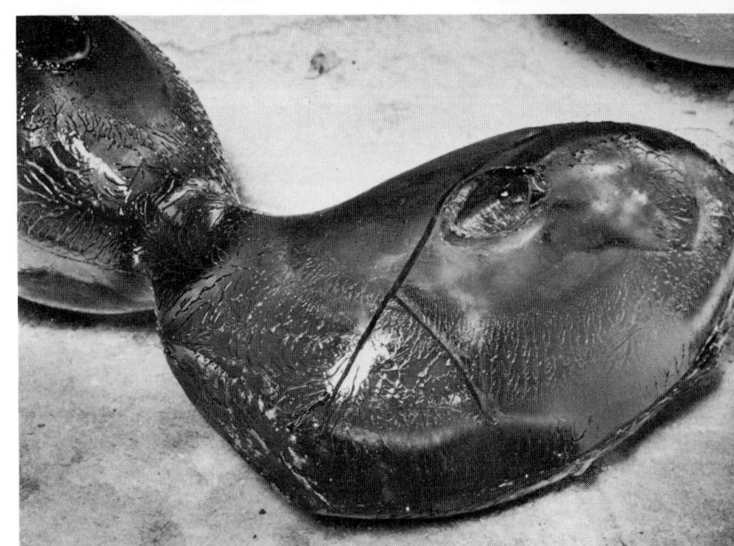

The crack is the result of an uncontrolled exotherm; either excessive heat was generated by a rapid cure and poor heat dissipation, or successive layers were poured before the previous ones had cooled.

Good casting as it comes out of mold.

Another form: a large finished plaster piece is prepared for an RTV mold.

Imperfections are patched with plasteline clay.

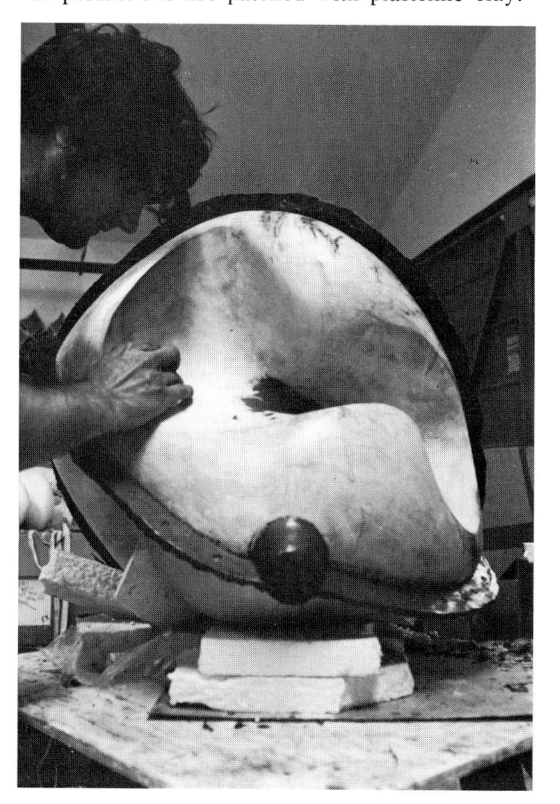

Coat with PVA and place on rigid foam blocks.

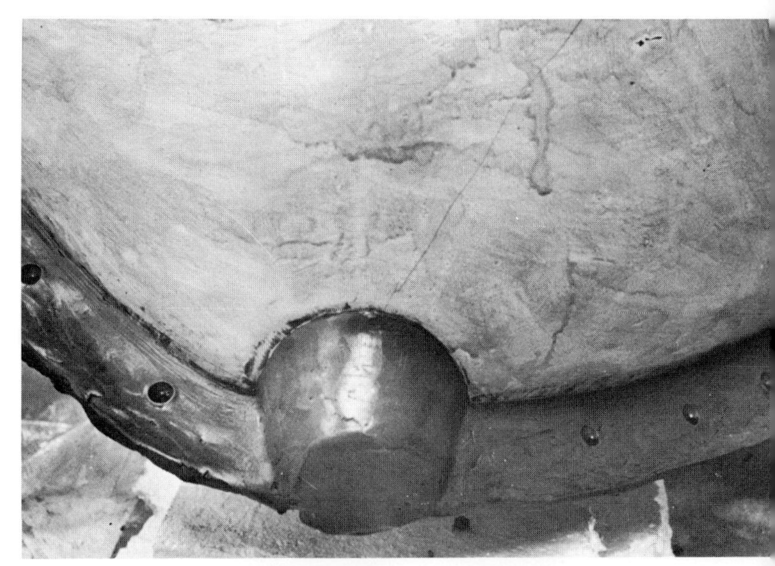

Detail of clay plug that will be pouring hole.

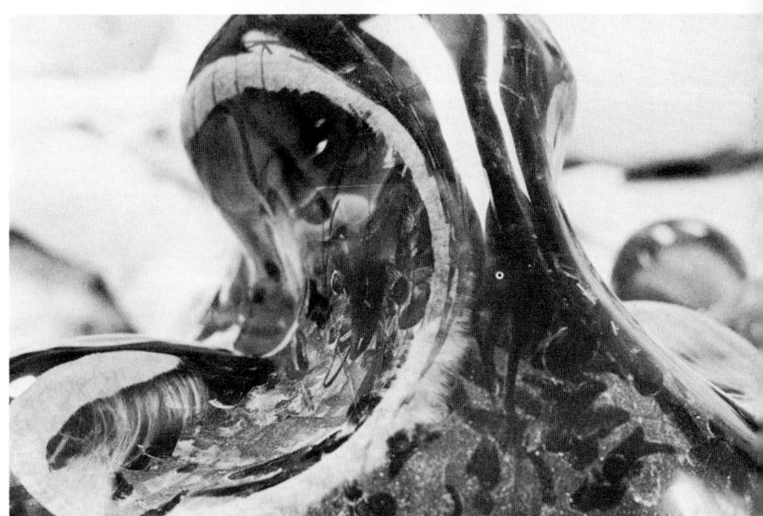

Detail of form changes on large casting by sanding and filing.

Finished multicolored translucent polyester casting by Roger Kotoske.

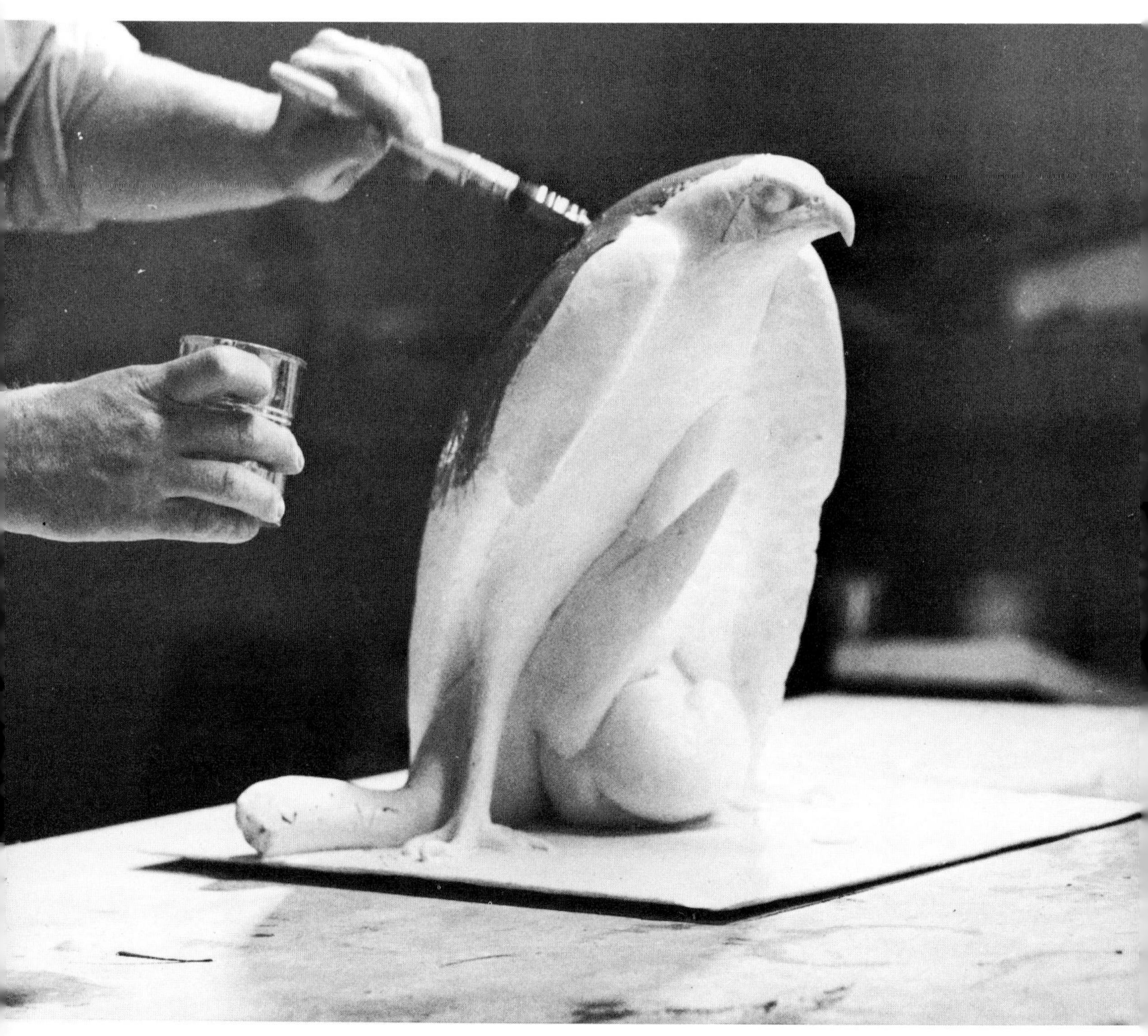

Epoxy pattern cast in plaster waste mold taken from clay original. First coat of CMC is being applied. Series demonstrated by Peter Fagan.

3

Cold Molding Compound (CMC): A Flexible Mold System

COLD molding compound was developed for industrial use in the late 1950s primarily by the Perma-Flex Mold Company. It was quickly abbreviated to CMC and just as quickly adopted by sculptors for casting sculpture from flexible molds. Black-Tufy CMC, the most familiar flexible mold material for many years, replaced gelatin and latex, especially where time was a consideration and expense was not. For most applications and large pieces, CMC is expensive but less so than RTV silicone rubber molds. CMC is dimensionally stable for an indefinite period. It cures without heat. A CMC mold may be taken from virtually any material, including itself. Some materials do require a parting agent or mold release.

CMC is a polysulphide liquid that cures at room temperatures by the addition of a two- or three-part catalyst system. The two-part system results in a mold of greater elongation than the three-part system, but its storage life in the uncured state is shorter. The three-part system has a more stable shelf life and its properties may be made more variable according to the various component ratios.

It should be observed that cold molding compounds may also refer to various silicone rubber mold materials and polyurethane resin base systems. As new materials and systems are developed definitions will vary. It is essential, therefore, to read and follow manufacturer's directions. Usually, the principles for use remain the same with minor variations in formula and procedures.

Sculptor Peter Fagan's demonstration of the use of a three-part CMC component system follows.

Left: To mix components for Perma-Flex (a trade name for cold molding compound, referred to as CMC) you will need: mixing sticks, scale, cans, rags or paper towels, and solvents. Mix thoroughly, by weight, and scrape sides of can. This is especially critical for small batches, less so for large amounts. Mix as little air into mold material as possible.

Below: The original epoxy casting may be altered and forms reshaped with a mixture of epoxy, Cab-O-Sil, and pigment to match the original.

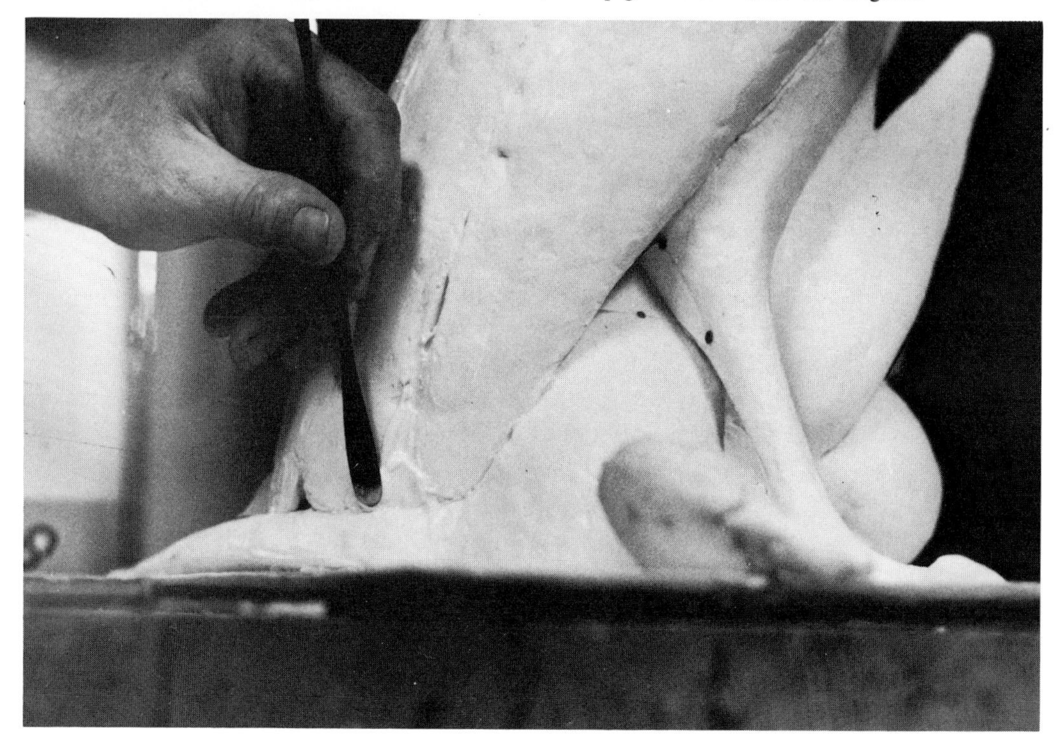

Right: Seams and crevices are cleaned and patched with epoxy mixture.

A pneumatic die grinder mounted with a carbide rotary file is used to trim or reshape epoxy.

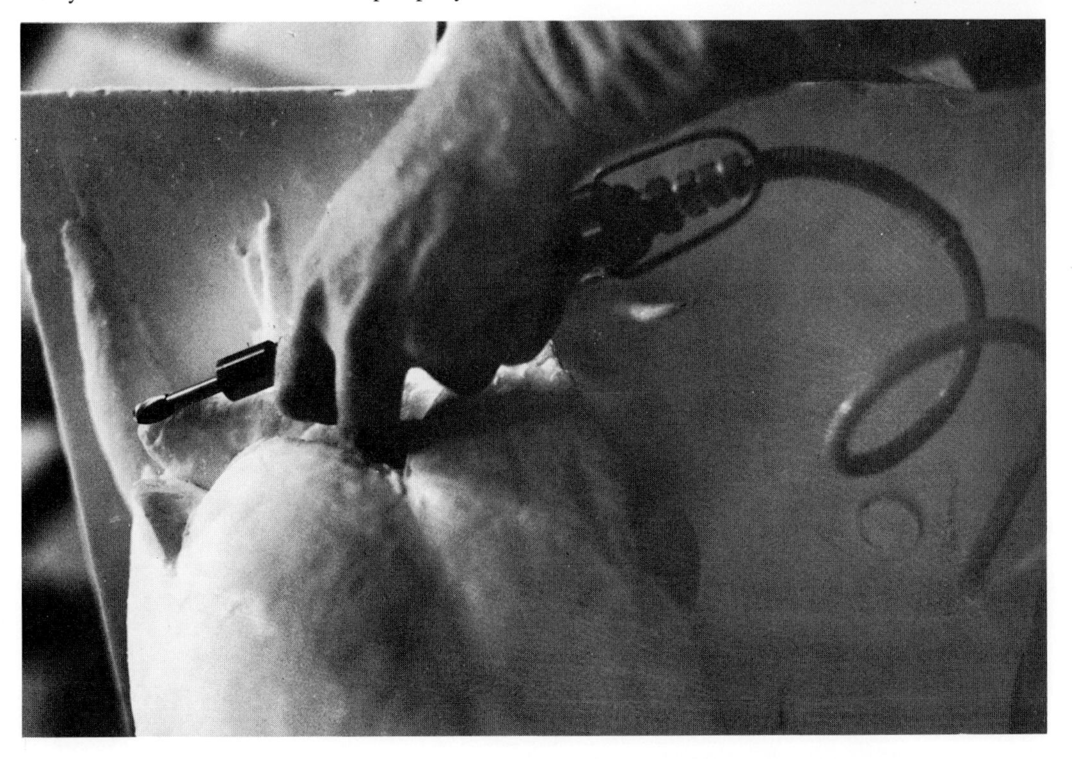

The first coat of CMC can be brushed on, but must be applied quickly as the gel time averages two to three minutes. Watch for bubbles that may later show up on the mold surface and use an air hose to blow them out; but use it gently as too much air pressure can add rather than remove bubbles. (Hot-melt room temperature setting vinyl, Blue Sil and Black-Tufy, also can be brushed, smeared, or poured. The method of application depends on your needs, experience, and the material you are working with.)

As you apply the CMC, if possible, place the piece at different angles so the compound penetrates into all undercuts.

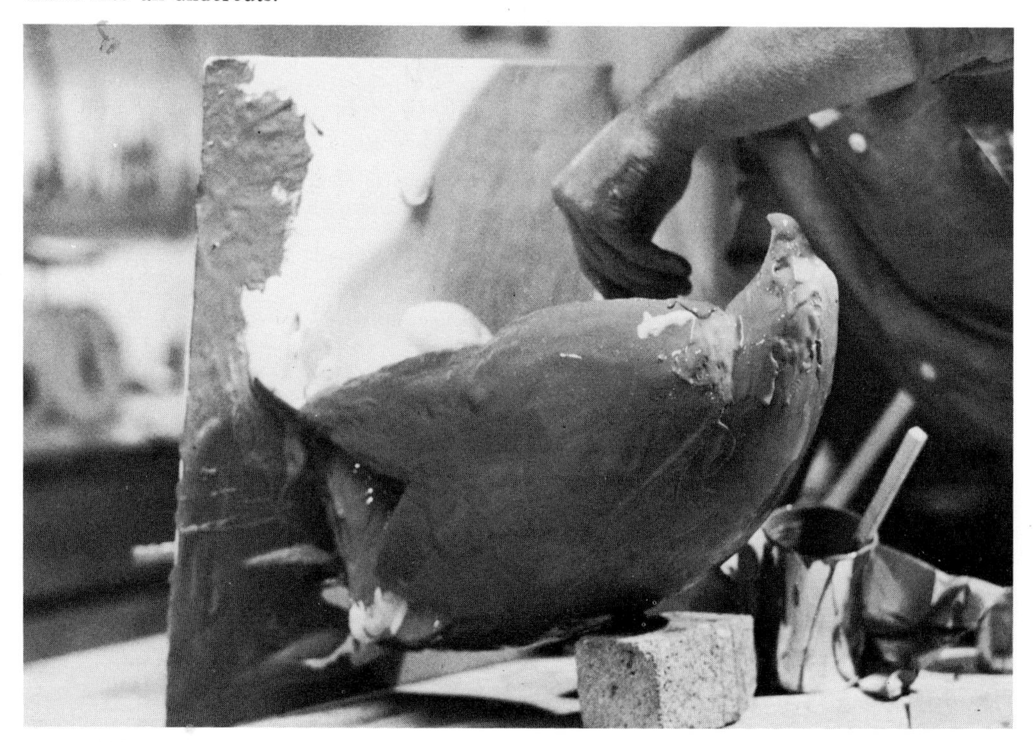

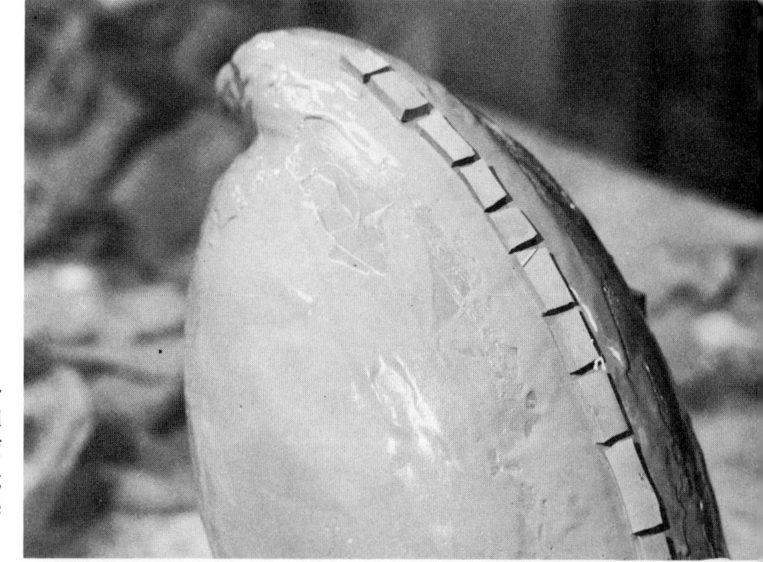

Scraps of set CMC can be used for various improvisations. Here they are used to build up thick areas along the back of the piece that will later be cut forming a thick seam, which will also allow the CMC to be removed from the casting.

Scraps may also be used as cushions when the piece is turned out.

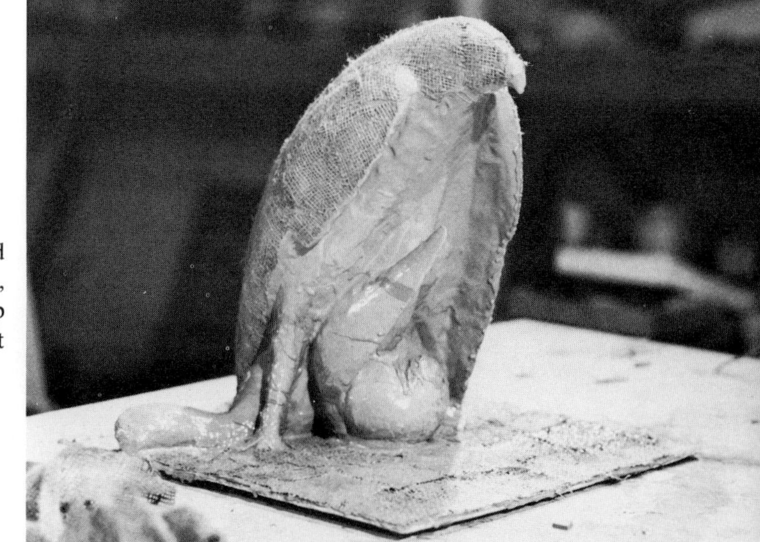

After the first coats of CMC are applied approximately ⅛ inch to ¼ inch thick, a reinforcement material such as burlap or gauze may be added to the unset CMC and coated again with CMC.

All undercuts are filled. Keys are added where needed, such as large, flat areas or sections that might fall out of register because of gravity or positioning. Keys also help prevent distortion as the weight of the cast material is applied to mold. (Note: The six tabs shown in the photograph later proved to hinder rather than help register CMC and removal of CMC from plaster support mold—they were cut down.)

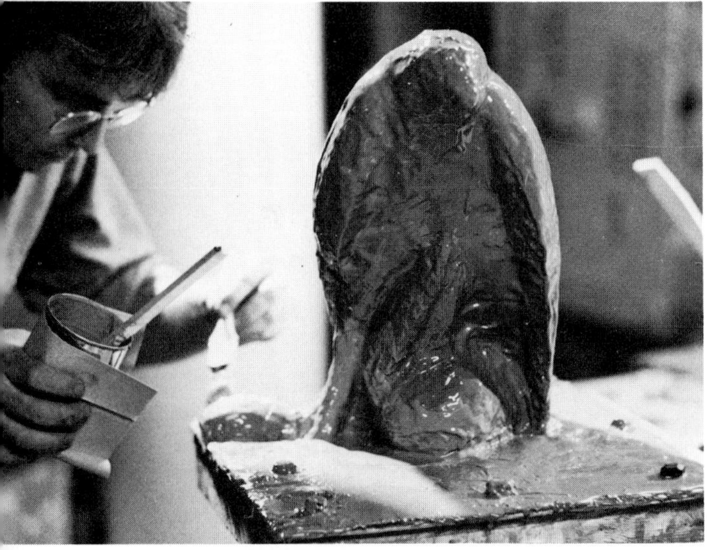

After the seams, keys, and reinforcement are covered over smoothly, allow the CMC to cure overnight.

PREPARATION FOR MAKING PLASTER CMC SUPPORT MOLD

Determine where the plaster support mold sections will be and how they will fit together and come apart. Build clay walls to define the first plaster section, which will be the basis for the remaining sections.

If undercuts remain, and the mold is stable enough, you can fill the undercuts with water clay, rather than make a small plaster piece mold section. This is an important consideration because small sections have to be made before large ones, and they are usually held in place inside the larger sections.

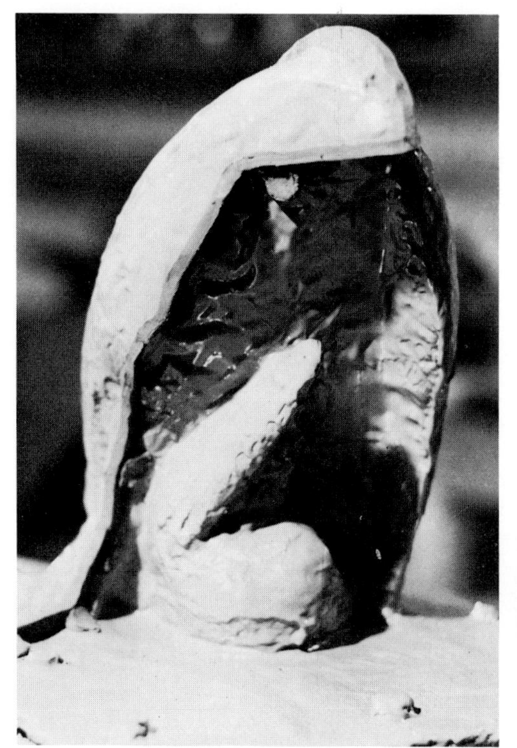

Front view of water clay dividers are seen at left.

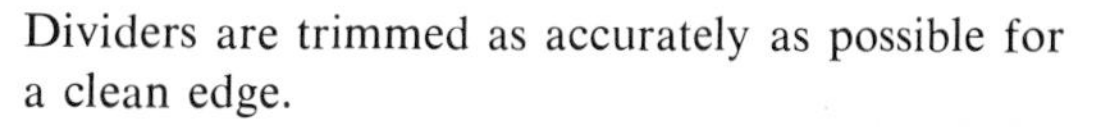

Dividers are trimmed as accurately as possible for a clean edge.

After the clay dividers are in place, coat the entire piece with green soap.

The first section of the plaster piece mold is laid up with plaster-saturated burlap as a reinforcement.

The section is then coated to an even thickness and areas of stress filled with plaster for additional strength.

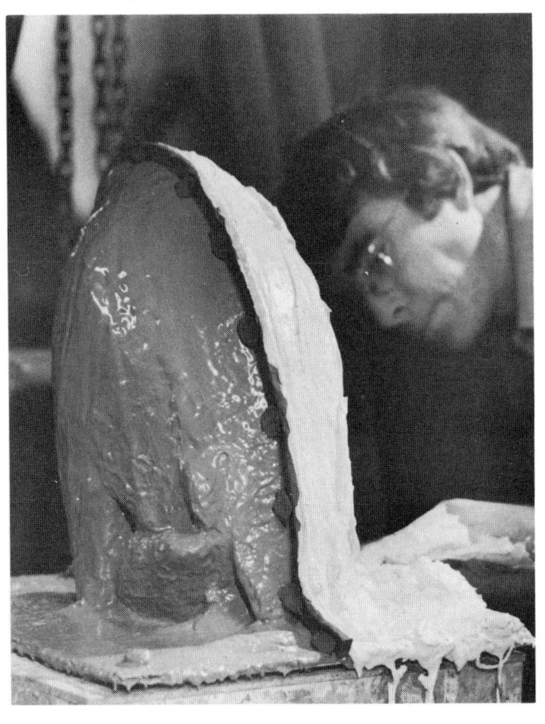

When the first section of plaster has set, the clay dividers are removed.

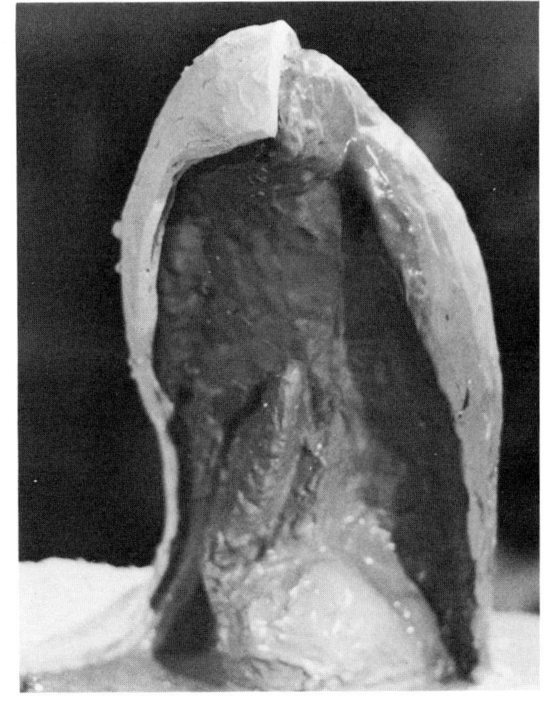

The edges of the clay and plaster walls around the final section are all angled out; the last section must be designed to lift out of the surrounding sections. Turn the mold on its back or side, when possible, to give you more control in placing plaster.

After mold is completed, gently separate sections. If sections stick stubbornly, release by pouring water down the cracks or use a blast of air.

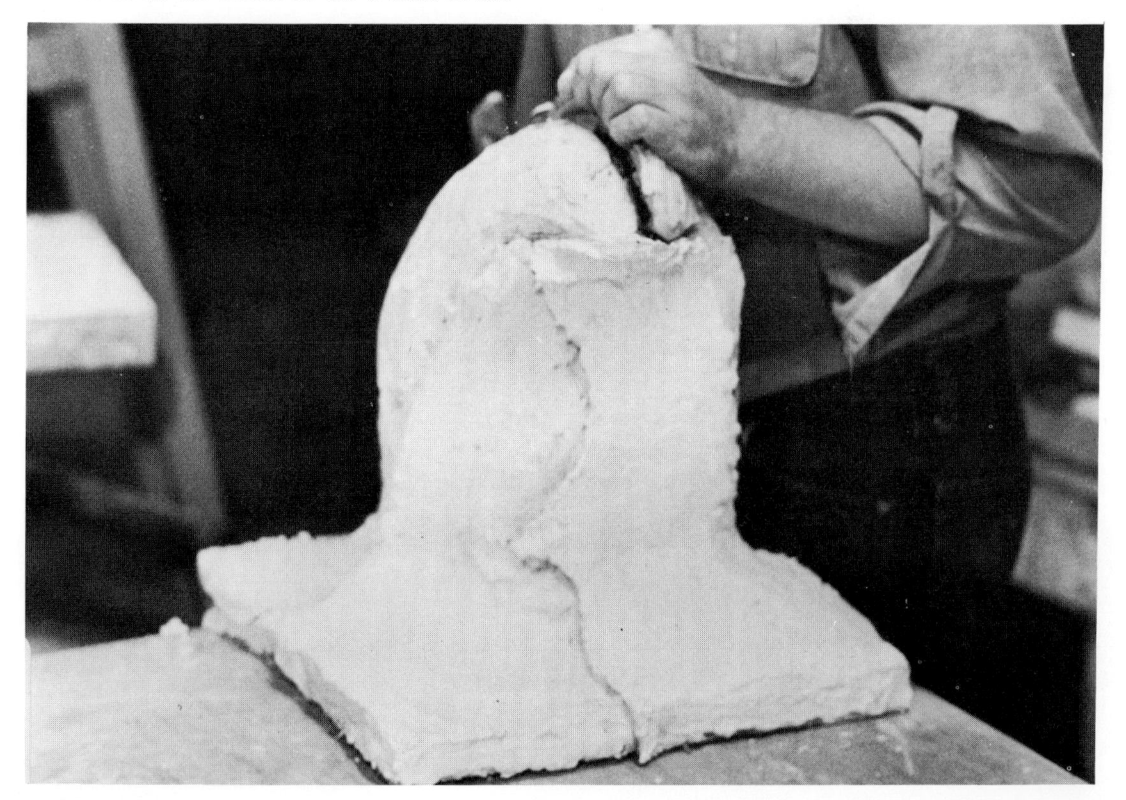

With the plaster piece mold removed, carefully cut the seams of CMC with a mat knife and again gently remove CMC from the original.

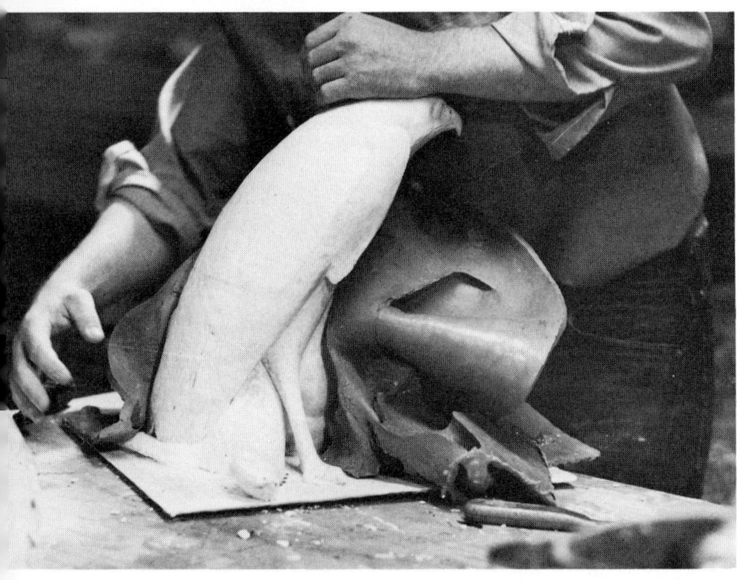

The CMC mold should pull easily away from the original. Be careful not to tear it.

Clean clay from support mold and trim surplus plaster from the edges with plaster knife. Sections are held in place with sections of inner tubes.

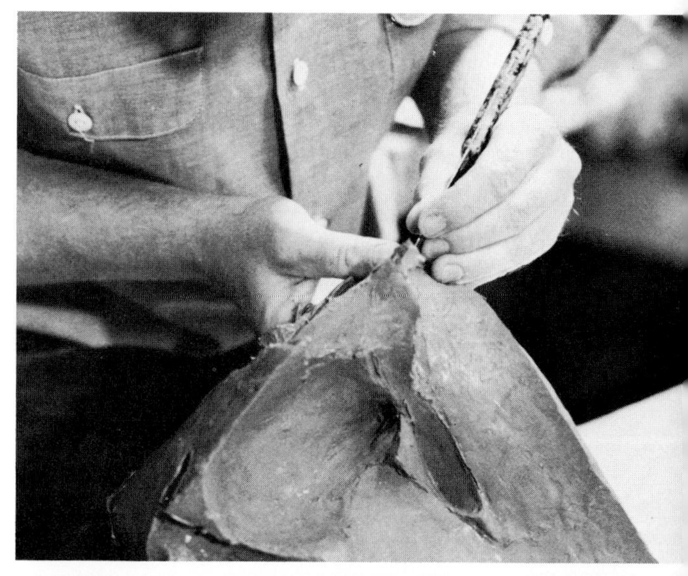

Clean the CMC mold and check all undercuts.

Reassemble the flexible CMC and support mold, secure it with cut inner tubes, rope, spring metal clamps, or wire.

The mold is now ready to receive the casting material. If wax is cast into the mold, the piece may be invested while still contained.

4

Latex Flexible Molds

LATEX, which is actually liquid rubber, offers many advantages as a mold material for sculptors. It is pliable, elastic, and more durable than any other flexible mold for the money. It is possible to safely and easily take exact impressions from originals with high projections, deep, complex undercuts, and to make scores of positives from the same negative rubber mold when fairly large editions of a work are needed in polyester, epoxy, wax, plaster, concrete, or latex into itself.

Latex is generally available ready to use. Various fillers, colors, retarders, or accelerators may be added for altering the basic qualities of the prepared latex.

To make a latex mold, and cast, the following general procedures are suggested.

The piece to be reproduced can be almost any material—clay, wood, metal, and so on. If plasteline clay is used it must be covered with lacquer or Krylon to protect the latex from the oil in the clay. Lacquer may be used to simulate textures in many places.

The piece is then placed on a flat, level, smooth surface such as Masonite. The level surface also serves as a support for the lip that goes around the mold which is usually the opening of the finished mold.

The first coat of latex is applied by brushing or spraying. All detail in the casting is dependent upon the first coat, so latex must cover all the areas you want to reproduce and also be free from air bubbles. The second layer, and the third, are almost equally important because each layer of latex adheres to the preceding one. Unless you are careful, the first layer may be pulled up from the original when subsequent layers are applied. Also, semidry latex is very sticky, especially when it is brought in contact with other latex. The lip that goes around the mold is also started at this time and it should extend about two to three inches around the piece.

After the first three or four coats are applied, you can paint on thicker latex layers until the mold is approximately ⅛ inch thick. (In the demonstration, there are between ten and twenty coats of latex on the original, allowing for two to four hours' drying time between each coat, depending upon the latex and weather conditions.)

Undercuts and protrusions can be made thicker or thinner to allow the reproductions to be removed from the mold more easily and to reduce broken parts. After the latex has been built up to the desired thickness and dried, allow three to four days for it to cure; this stops excessive shrinkage and distortion and permits thicker areas to dry. Coat the dry latex with shellac or wax.

Next, a support mold is built and, depending upon the undercuts and protrusions, it can be made of one or several pieces. The most basic support mold is plaster. Polyester or epoxy reinforced with fiberglass is also used for a lightweight support mold. The finished support mold reproduces the exact form of the latex and gives it support for distortionless casting. After the support material has set, it can be removed from the latex. The latex is then removed from the original piece.

To retain a hole through a piece to be reproduced, place a thin metal or cardboard shim in the center of the hole. Paint the latex from either side of the hole covering the shim. The latex can then be pulled from each side of the shim and, when the piece is cast, the thin sliver of cast material that takes the place of the removed shim can be cut away.

Pieces with large undercuts may require support molds that come apart to allow the casting to be removed from the mold. If undercuts are small enough on the reverse side of the latex they may be filled with clay and then the plaster support mold is made over the latex and the clay. Usually, when many smaller pieces are required, they are all held in place by another large support mold. To remove the piece, the mold is turned over, the overall support mold is

OMINOUS IKON. Dennis Kowal. 1968–70. Epoxy and fiberglass, 8 feet high. Krannert Performing Arts Center, Champaign, Ill. This is the finished sculpture of the process illustrated in the following demonstration.

removed, then the small pieces and finally the latex taken away. This type of casting can become complicated, but usually a little common sense will provide a solution. Always remember that each piece to be removed must fit tightly and be removable in turn, as each piece is usually held in place by another.

When making the separate support mold parts, always shellac the plaster surfaces and use a parting compound such as castor oil, paste wax, or green soap. If not coated, the plaster surfaces will stick to each other.

To clean brushes used with latex, wash out in water and dip in lacquer thinner, then brush out particles of latex with a wire brush.

LATEX MOLD FOR EPOXY CASTING

The surface of the original plaster piece has been sealed with shellac. A first coat of latex, as it comes from the can, is applied very carefully with a large brush. The first coat captures all details, so bubbles are to be avoided. If bubbles do appear, blow them out using a light pressure air stream.

Here the piece has approximately four coats of latex. Ten to twenty coats will be applied, allowing about a two-hour drying time between coats. A fan circulates air at all times. Drying between coats is vital as shrinkage will increase if each coat is not dry. Humidity and temperature and the amount of latex in the ammonia-water solution affect drying time.

After approximately fifteen coats of liquid latex, burlap reinforcement is laid directly on a wet coat. Burlap pieces should not overlap except at the seams: this allows latex to flex without tearing.

Burlap completed, coated three times. The last coat is still wet.

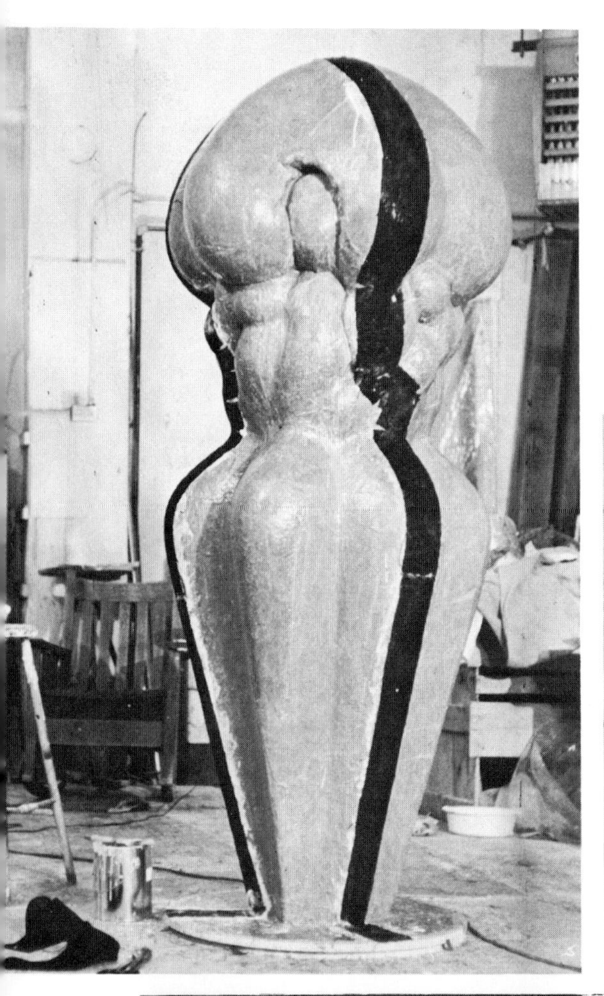

Seams that will be cut after the flexible mold and support mold are completed and reinforced with double layers of burlap and coated with two additional layers of liquid latex. Black burlap is used to define seams.

A temporary enclosure was built from plastic sheeting and wood stripping. A plastic tube was run to the opening of an overhead space heater. The trapped heat was directed on the piece for three days to assure maximum drying and curing of the latex. This also reduces shrinkage, especially when cured on the original. An auxiliary electric space heater was also used.

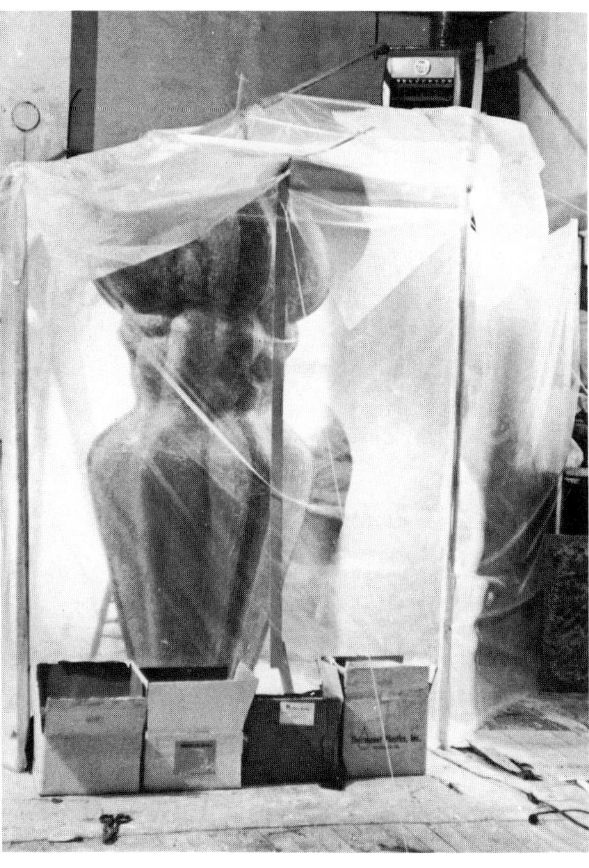

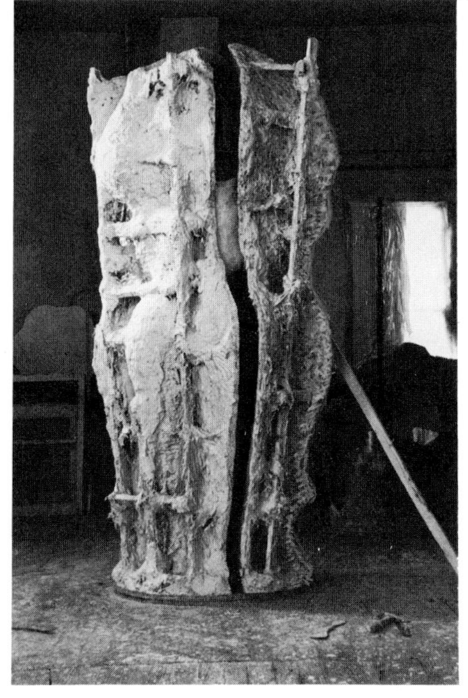

BUILDING THE SUPPORT MOLD

Clay dividers are then made directly on the latex surface to define the first of three plaster support mold sections. After the dividers were made the latex was waxed to facilitate removal of the sections when all are completed. Sisal fiber, used as reinforcement, was dipped in plaster and laid up on the latex. This results in a light, strong, thin support mold.

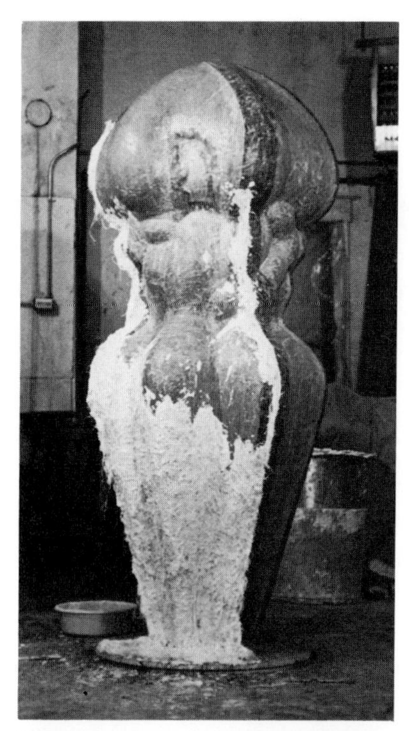

Wood strips are used for reinforcement of the completed first section. Clay dividers are removed and a new one laid to define the second section of the mold. The edges of the first section are waxed to prevent other sections from sticking.

Two sections of the mold are completed. Seen through latex are extra thicknesses of burlap at seam areas. Each section, because of the extensive use of sisal fiber, weighed approximately 100 pounds.

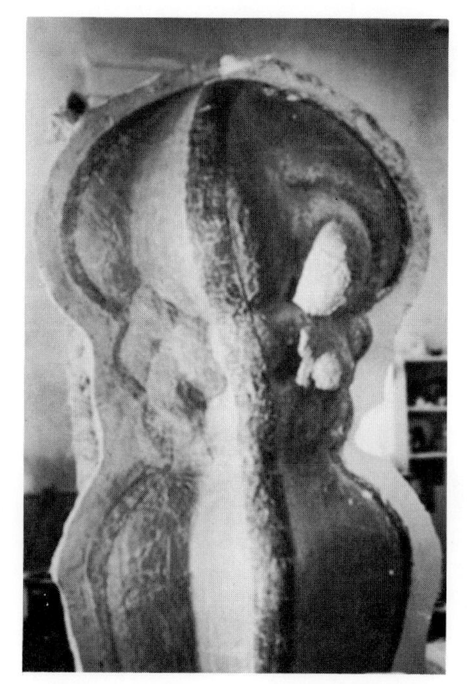

Photos of the completed mold partially open. The remaining sections are removed; the latex mold is cut along predetermined seams that match and fit the support mold.

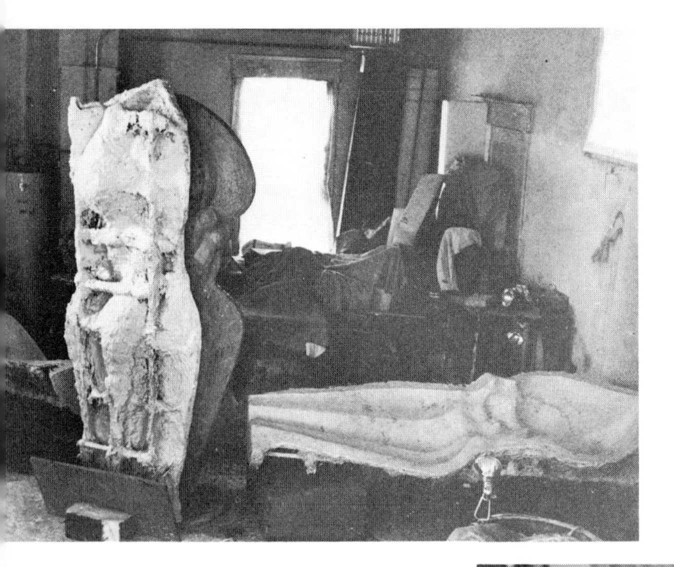

CASTING EPOXY

After cutting the seams, the latex is removed and fitted into the support molds, cleaned, and all surfaces inside and out are coated with talc to prevent the mold from sticking to itself. Fasten to the plaster if necessary to achieve accuracy, and coat with wax, since epoxy is to be cast into the mold. Wax, epoxy, plaster, and polyester may be cast into latex, but always use the appropriate mold release (wax, silicone, polyvinyl alcohol, etc.). Check register of the flexible mold to support mold as a mis-register could occur. (See chapter 5 for another epoxy casting series.)

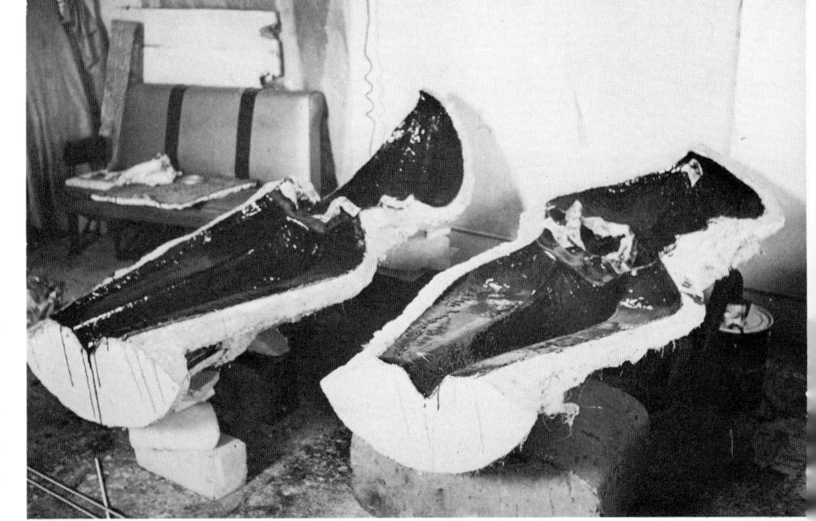

Two sections of plaster support mold and latex mold shown with three coats of *brushed-on* epoxy. The third section (not shown) is the same. The epoxy mix is made of one part resin, one part catalyst, and powdered ceramic glaze (burnt umber) for color. It is mixed with an electric drill and mixing attachment. The resin and catalyst were mixed, then the color.

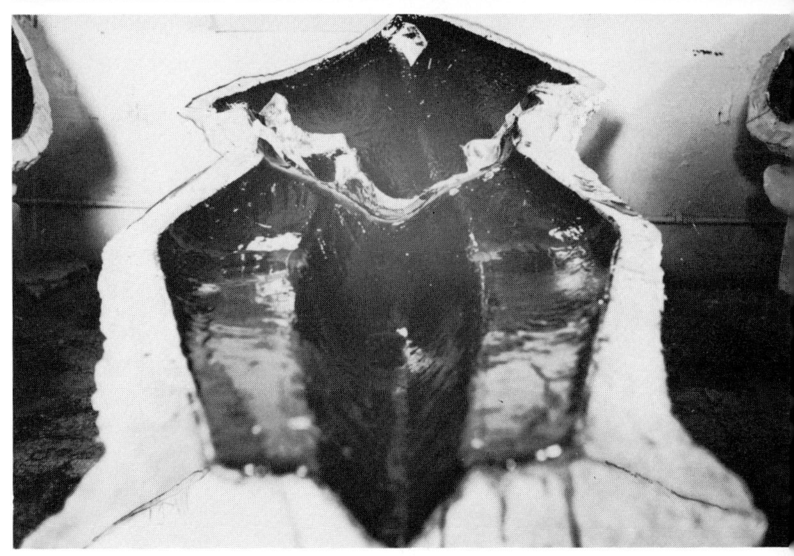

Detail of one section of the mold. After the initial coatings of epoxy, the remaining ones were heated, causing a fast cure. A different catalyst or more of it will also accelerate curing time.

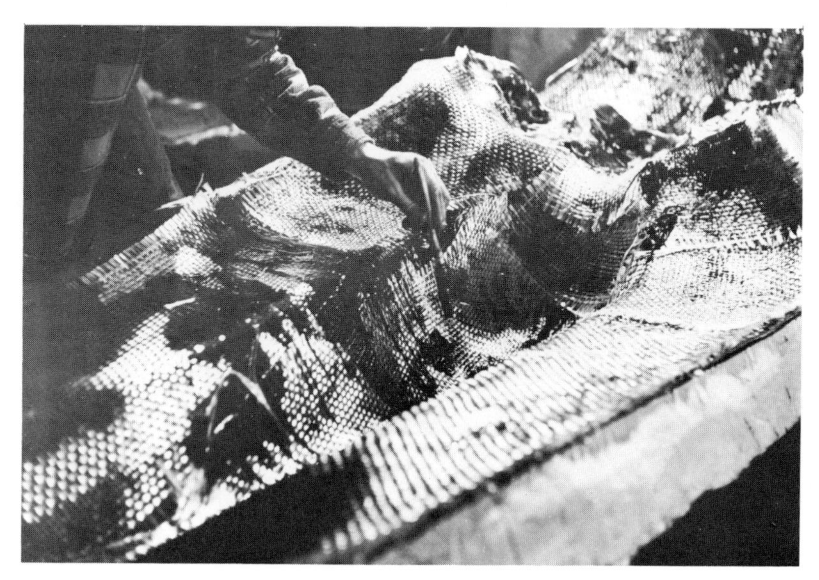

Fiberglass cloth is pressed into the unset epoxy. Two grades of fiberglass cloth were used: finer cloth for detailed areas and heavy cloth for broad areas. Double layers are used at stress points.

Trim excess protruding fiberglass and epoxy from seams so they will register accurately. Shown, a sanding drum with a 50-grit cloth-backed sleeve on a high-speed (24,000 rpm) die grinder. It is imperative that goggles and respirator be worn during power sanding.

Sections are aligned. Thick seam edges were made by mixing cotton flock with epoxy for a very thick mix. Glueing the sections was made easier by this operation.

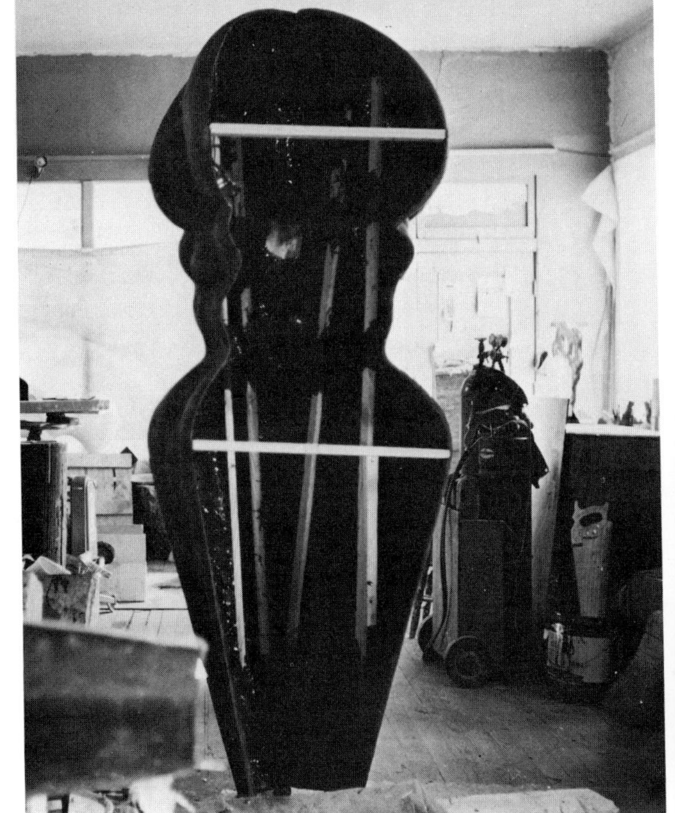

Two sections are shown joined. Wood strips, held in place with epoxy and fiberglass, are used for lifting and reinforcement.

Eve. Tom McClure. Bronze and epoxy.

Breaker. David Black. 1967. Aluminum.

Tribute to Those Magnificent men and Their Flying Machines. J. Fred Woell. 1969. Epoxy.

Untitled. Roger Kotoske. 1970. Polyester resin.

Forked Hybrid. Richard Hunt. Bronze.

Untitled. Morris Applebaum. 1970.
Epoxy and vinyl.

Ashford Medical Building. Puerto Rico.
Rolando Lopez Dirube. Cast concrete.

American Woman. David Hostetler.
1968. Polished bronze.

Seams are prepared for matching and filling by running masking tape on either side of seam about ¼ inch from the edge. The distance depends on the size of the piece and whether or not the seam edges match accurately. Liquid epoxy is poured into seams, allowed to set, and filed to blend with the surrounding areas. All seams were reinforced with fiberglass from the inside of the mold, which also prevented the epoxy, poured from the outside, from running through.

BIG RUBBER LADY. Eldon Danhausen. 1969. Latex, approximately 40 inches in diameter. Piece was modeled in clay and a two-piece plaster mold made of the clay figure. The plaster mold was removed, washed, and allowed to dry thoroughly. A separator of two coats tincture of green soap was brushed into the plaster mold. About 25 coats of rubber, yielding about a ¼-inch thickness, were slowly brushed in. Coarse burlap was laid in between the last few coats of rubber for reinforcement. Plaster mold was broken away to reveal rubber figure.

Courtesy, artist

MERCATOR'S P. Paula Tavins, 1969. Marbleized latex pieces cast from a plaster negative mold taken from the clay original, 12 inches high, 18 inches wide.

Betty Parsons Gallery, New York

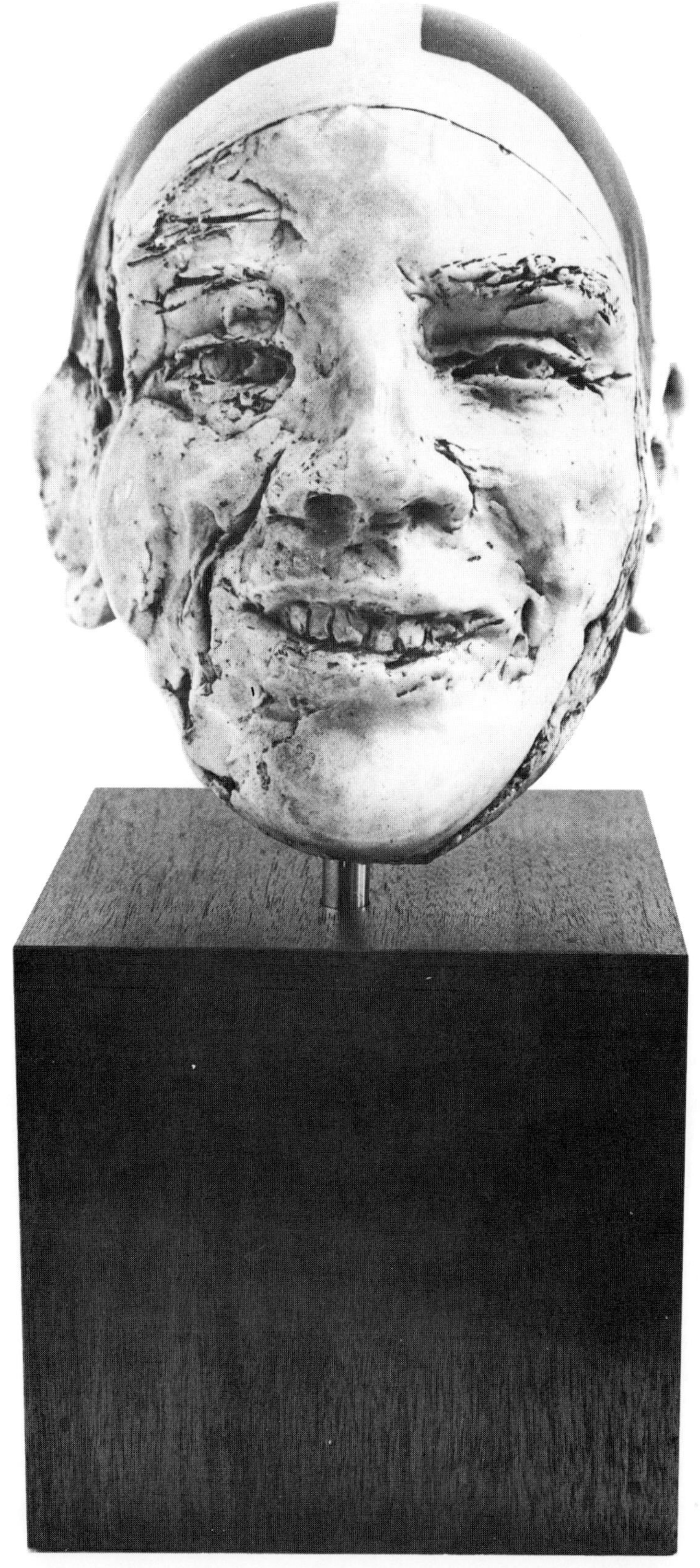

SCUBA HEAD. Fred Woell. 1969. Cast epoxy resin, 12½ inches high, 5½ inches wide, 8 inches deep.

Courtesy, artist

5

Casting Epoxy and Polyester Resin (Plastics)

BOTH epoxy and polyester resins have been used industrially for many years to cast products, parts, adhere parts together, as protective coatings, and to make fibers for textiles. As the cost of high-quality plastic has gone down, plastics have become more commonly used. For the sculptor this is a great advantage. Plastics now offer high quality, low cost, flexibility for various casting techniques, impregnation, an infinite color range, and transparency, translucency, or opacity.

To cast plastics into molds, the following general procedure is recommended: Read and understand manufacturer's technical information sheet; be prepared with all containers and tools; have mold clean, dry, and coated with separator; mix components; apply first (gel) coat by pouring or brushing into mold; check for air bubbles; let cure. If fiberglass is used, apply another coat of resin over the first coat; when tacky, press in fiberglass and apply a third coat of resin.

When a separator or mold release such as PVA (polyvinyl alcohol) is used on the mold surface, which is a delicate film-type coating, care must be taken not to break the film. Wax separators are more durable and cover less detail. The wax is thinned with carbon tetrachloride or white gasoline, half and half; buff the surface when the wax dries. Wax may also be melted and painted into the mold.

To accelerate the curing of the first coat of resin the mold may be heated, or the resin heated before the catalyst is added, which also thins it out, allowing bubbles to rise to the surface before the catalyst is mixed in. If air bubbles are observed, they may be removed by a blast of clean, dry air or by quickly passing a neutral oxygen acetylene flame over the surface.

FILLERS

Filler materials may be used to lighten a casting. Such fillers are Perlite, phenolic, and polyvinylidene Microballoons, pecan or walnut shells, flour, aluminum pellets, cotton flock, Cab-O-Sil, aluminum silicate, or calcium carbonate. Some fillers such as the nutshells or aluminum pellets are used to absorb the heat generated in large castings. Fillers are added after the resin and catalyst are properly mixed.

REINFORCEMENTS

Depending upon the thickness or strength of the casting desired, one to six coats of plastic may be applied. After one of the coats is applied and, before it sets, fiberglass mat cloth or chopped fiber may be added for strength and extra material placed at stress points. The reinforcement material is pressed into the resin. A second coat of resin may be immediately applied after the fiberglass is laid or after the preceding coat has partially set; this creates a good bond between the reinforcement and the resin.

Metal reinforcements and fasteners such as steel reinforcement rods, wire, nuts, bolts, or hooks can be added at this point. However, with plastics, additions or changes may be made at any time.

COLOR

Color is added as powdered pigment or liquid dye after the resin is mixed and can be varied from opaque to transparent. Individual colored areas may be cast by masking off determined portions with tape or clay in the mold. After casting, the masking is removed and the remainder of the casting done up against the colored areas or objects to be impregnated.

If sections cast separately are to be matched, they may be joined while still in the mold or after removal. Flanges should be made along all seam areas while the pieces are still in the mold to provide a good bonding area between sections.

FILLING SEAMS

Individual sections are simply glued together and then the seams are filled. To fill seams, tape approximately ¼ inch on each side of seam and fill void with resin. In most cases you will be able to get at the inside of the seam to apply fiberglass reinforcement. After the resin has set, remove tape and file or sand the area around the seam to blend with surrounding area. Bubbles or blemishes may be drilled out and filled.

SANDING, POLISHING, AND FINISHING

Almost any file will work on plastic. The result desired will determine the file to use for the job. Multipurpose 4-in-1 files are excellent; flat and round hand files work well. Spiral-cut rotary files in an electric drill are good. Drum or disc sanding, using open-coat aluminum oxide, 150 grit, will remove material quickly. Hand sanding works best with wet/dry carborundum paper, 220 for fast smoothing. Most plastics can be easily hand polished after using 400 then 600 grit wet/dry paper. Machine buffing is most efficient using a commercial auto body rubbing compound. You must be careful to make sure the compound's color doesn't discolor the plastic and that you move over the surface quickly at a low rpm or else the plastic will burn.

For a polishing agent you can use aluminum oxide, pumice, or fine grits of rottenstone. A flame passed quickly across acrylic will both polish the surface and blend fine sanding or file marks. Avoid burning, which can cause stress marks or crazing on the plastic surface, especially on clear or translucent pieces.

Wax, lacquer, enamel, and acrylic coatings can be used on plastics. Some actually eat into the plastic surface, providing a permanent bond, while others are only a surface coating. An acetylene flame used on epoxy can change its color. The area to be burned must be masked with aluminum tape or heat-resistant tape. Always experiment on scrap.

If a flexible mold is being used to make multiple castings, removal of the piece is simple. However, if a plaster waste mold is used, removing the piece can be difficult and time consuming, especially if the correct mold release such as wax or silicone was not used or was not sufficient. The pounding required to break away the plaster can cause surface cracks and break off details. When using a plaster mold, work the main masses away first and gradually delve into detailed or weaker sections, always giving the piece ample support. If the mold is coated well with a release agent, large areas will pop off easily and you will have a clean casting.

SAFETY

Working with all plastics presents hazards, for their additives have harmful effects. Industry has been reluctant to clarify the extent of side effects, but serious disabilities have been reported in recent years. Therefore, always work with the best ventilation possible, wear a respirator and gloves, and avoid direct contact with the material. Try not to get it on yourself or anything else. Take special care to keep it away from your eyes. Keep materials and work areas clean. Always remember that many of the chemicals associated with plastics are explosive. Every precaution, combined with common sense, should be exercised.

In the following series, Tom Goldenberg has made a plaster mold from an original piece modeled in clay over a welded steel armature. Separated sections were designed for best removal from the clay, and, later, the epoxy casting, and so seams would join at inconspicuous places on the finished piece.

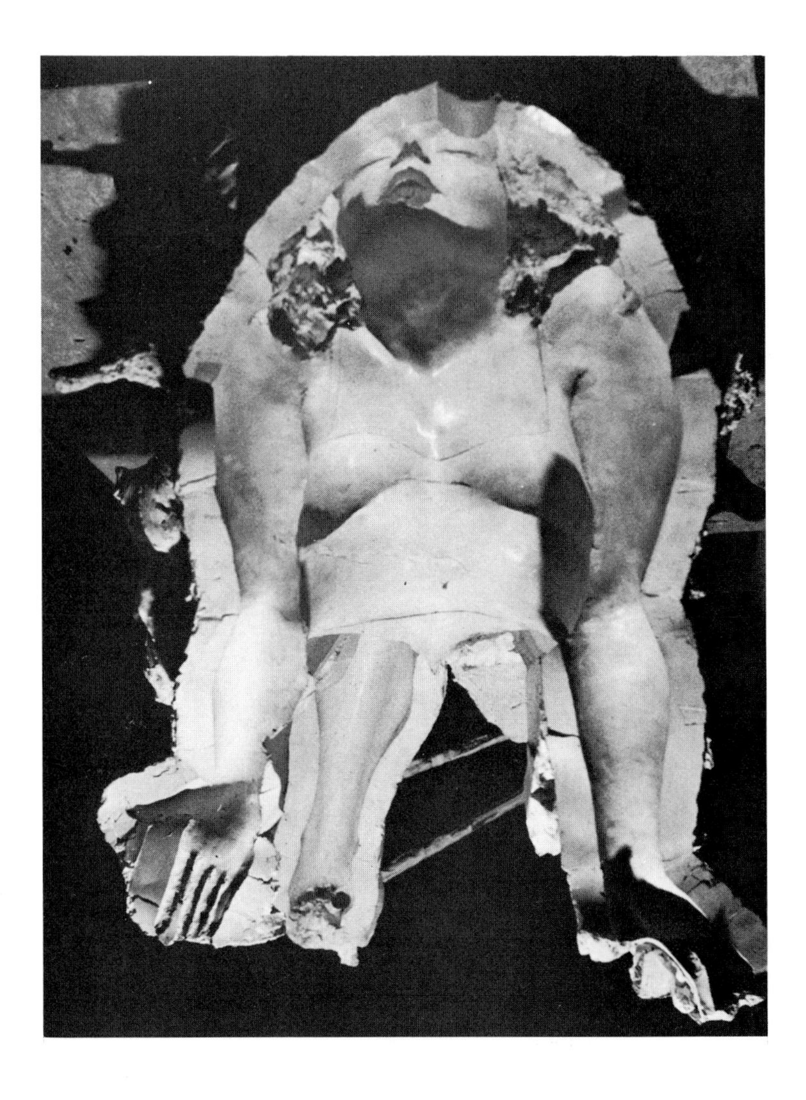

The plaster mold by Tom Goldenberg has been cleaned of all clay. The surface was sealed with a thin shellac coat and then a release agent such as Johnson's Paste Wax wiped in and polished out.

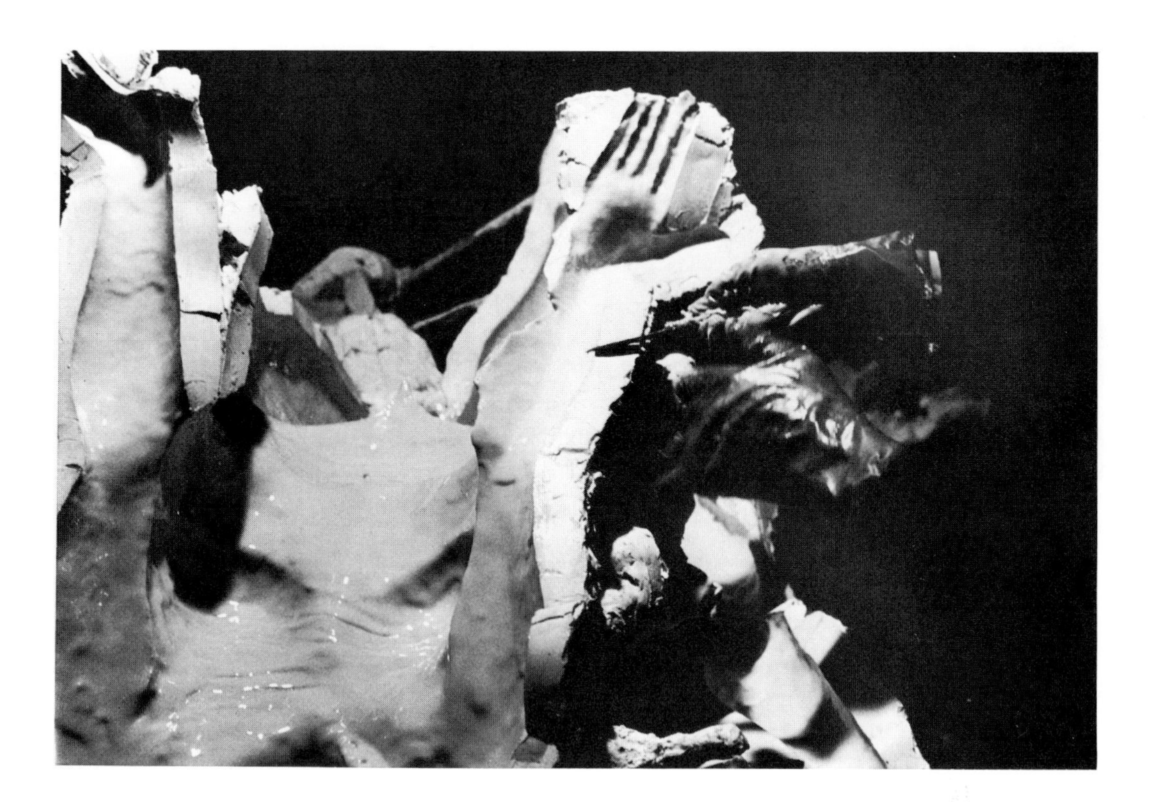

A small quantity of Cab-O-Sil was mixed with the epoxy to increase viscosity. White and yellow pigment were mixed together for color and the mold coated with the combined mixture of Cab-O-Sil, epoxy, and color. For protection, wear disposable plastic gloves.

Fiberglass cloth was then pressed into the second or third epoxy resin coat and allowed to set. Several dyed or lightly pigmented coatings may be applied to achieve translucent color, or an opaque layer may be laid using intense colors or fillers.

The fiberglass was *completely* coated with epoxy.

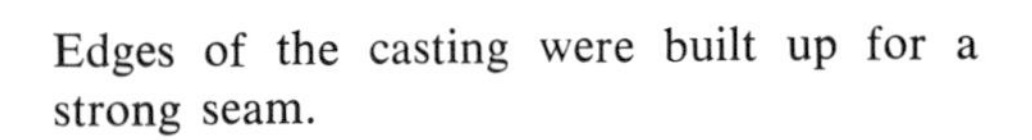

Edges of the casting were built up for a strong seam.

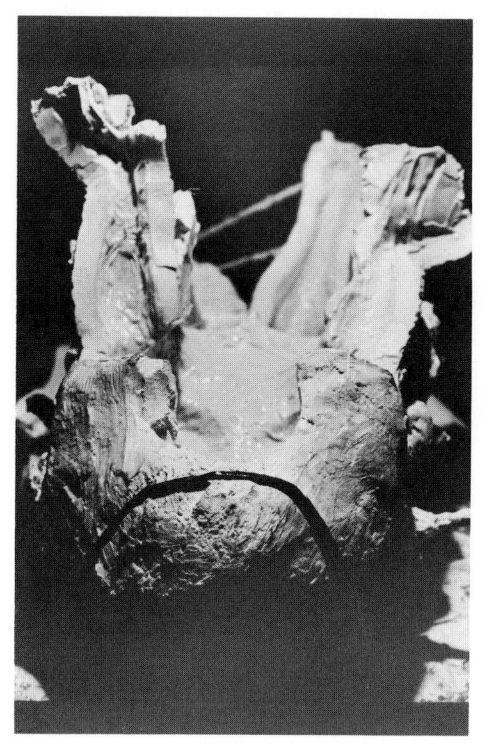

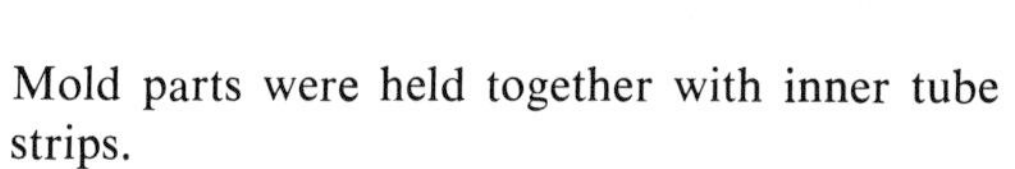

Mold parts were held together with inner tube strips.

While still in their respective molds, two sections of the casting, cast separately, were coated with epoxy and tied together with rope until set.

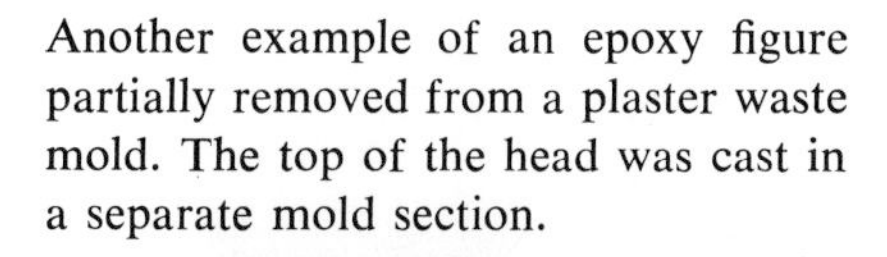

Another example of an epoxy figure partially removed from a plaster waste mold. The top of the head was cast in a separate mold section.

Detail of a seam showing some plaster still remaining, which must be chipped off.

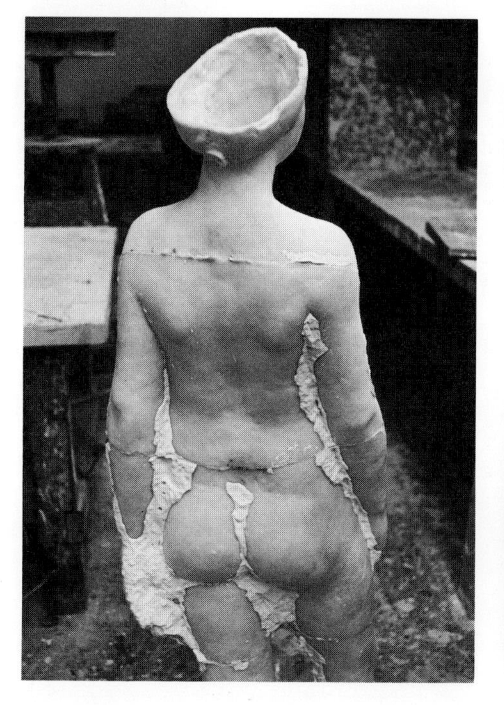

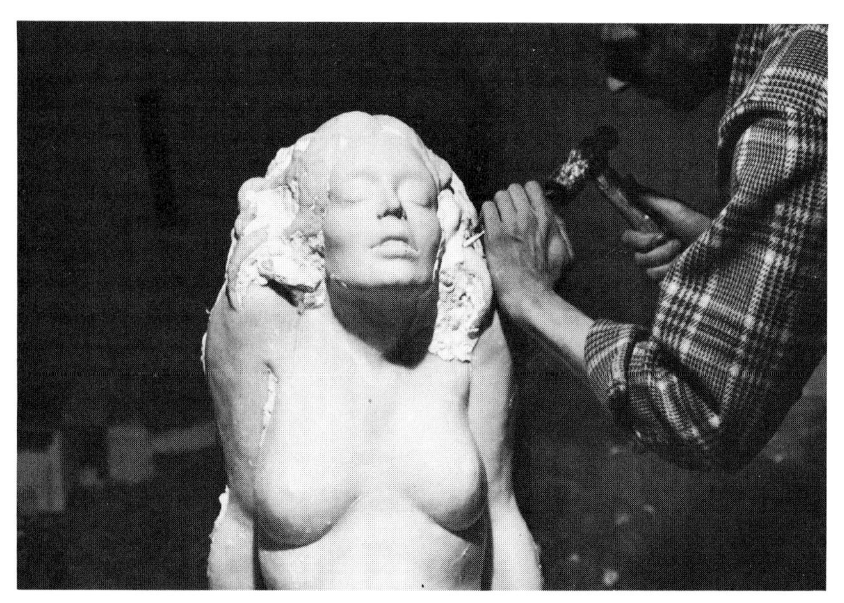

Top portion of assembled piece. Final chipping away of plaster in detail areas must be done very carefully.

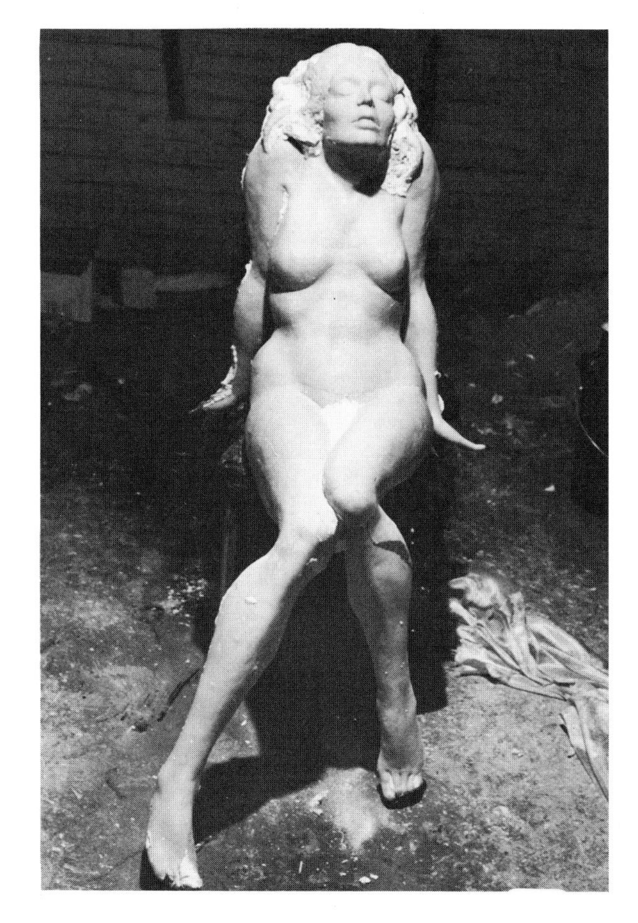

Piece as it appeared almost cleaned of plaster. Surface was then cleaned with alcohol or a solvent to remove wax residue. Seams were filled. Voids were filled with epoxy; some voids, too small to fill, had to be drilled out, then filled.

Finished casting. The surface quality desired determines the amount of sanding and polishing.

POLYURETHANE FOAM

Polyurethane foam, not to be confused with polystyrene (Styrofoam), has many characteristics that make it a good material for casting lightweight sculpture. Polyurethane foam is available as either a thermoplastic or thermosetting material, it can be foamed at a wide temperature range, and may be flexible or rigid. Its density can vary from 1 pound to 50 pounds per cubic foot and can also be developed with a thin skin or thick, such as Foam Hide by Teledyne.

Polyurethane foam components are very sensitive to moisture, and (B) components, for example, have a low boiling point, approximately 75° F., and must be stored in a cool dry area, especially before exposure to air. Opened containers can be purged of moisture by a blast of dry air or nitrogen which will not adversely affect the material.

Standard mixing procedures are followed: The isocyanate (A) component is first weighed and poured into a container for mixing; the polyol (B) component is added and blended. Use containers of paper, metal, polyethylene (which is easy to clean and reuse), or any non-contaminating material at least twice the size of the batch. During expansion the foam gives off toxic fumes.

Casting is usually done into silicone or urethane elastomer molds because they are easiest to make. A mold release must be used on all mold surfaces except for silicone rubber molds, but with difficult castings a release of some kind may be needed even with silicone molds. In some cases a warm mold can be used that will cause a thin skin to develop; conversely, a cool mold will cause a thick skin to develop. Temperature control of the (A) component will accordingly change the results.

It is always wise to follow manufacturer's directions and to experiment before proceeding with a casting. It is also important to remember that since the mold being cast into should be sealed or partially sealed, the use of vents should be considered along with how the mold is to be clamped together and how, if at all, the expanding foam might distort the mold. A reinforcing or support mold might be required.

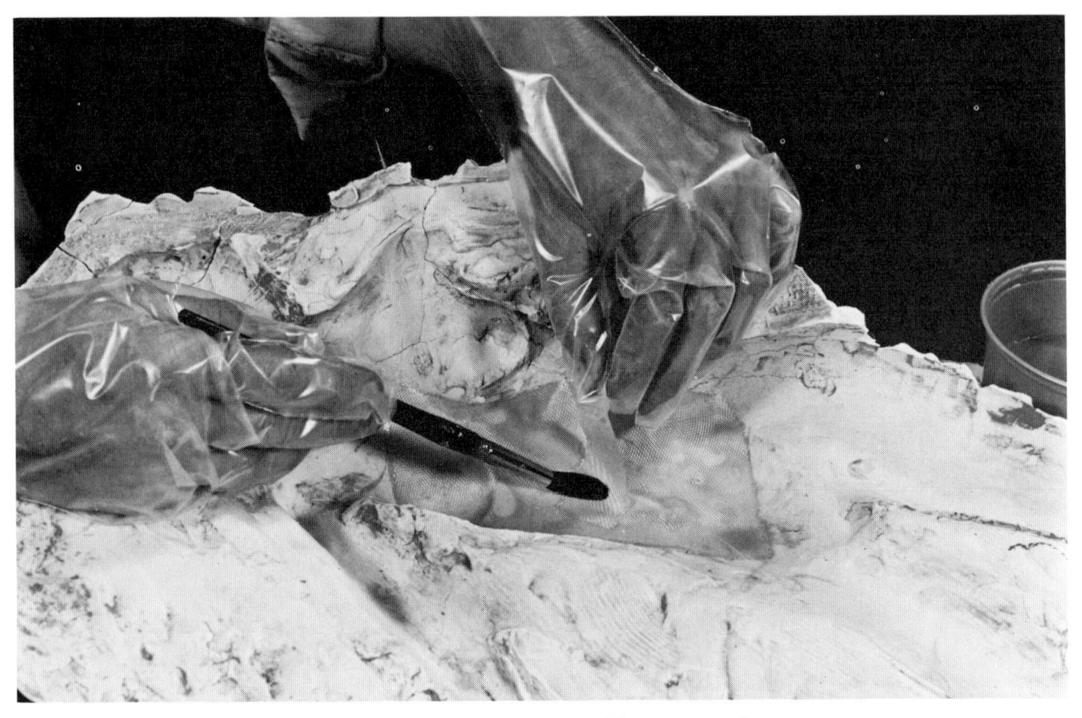

Fiberglass cloth is pressed into plaster mold coated with epoxy resin.
Courtesy, William Robertson

Spraying a vacuum-formed piece. A spray booth is used to draw out fumes. A respirator is worn to protect lungs.
Courtesy, William Robertson

WHITE I. Mac Whitney. Lucite sculpture.

Courtesy, Main Place Gallery, Dallas

CONSPIRACY. Arthur Secunda. Cast and laminated polyester resin assemblage epoxied to a black one-inch acrylic base, 20 inches high, 20 inches wide, 2 inches thick. Work was made by pouring ¼-inch coats of clear casting resin into a Plexiglas "box," arranging objects between pours. The final layer of resin (on rear, not visible in this photo) created a subtle, pinkish blush after repeated wet sanding and polishing. The final coating of resin contains wax, which surfaces upon curing, thus providing a durable and slicker surface for finish.

Courtesy, artist

I AM HE AS YOU ARE ME AND WE ARE ALL TOGETHER. David Schneider. 1960. Powdered iron in polyester resin, 22 inches high, 26 inches wide.

Courtesy, artist

SPAN. S. Porter. 1969. Fiberglass, 24¼ inches high, 98 inches long, 26 inches deep.
Betty Parsons Gallery, New York

Bruce Beasley with autoclave used in acrylic casting.

Bruce Beasley polishing the casting of sculpture KILLYBOFFIN. Pneumatic power tools and a special workbench facilitate final finishing.

Bruce Beasley's huge clear acrylic resin sculpture *Apolymon,* weighing six tons and measuring 15 feet, 8 inches long, was commissioned by the State of California. Beasley had to invent a process to cast acrylic resin in this size, and did so after typically being told by industry that it could not be done. He bought 55 gallons of methyl methacrylate and tried different catalysts and proportions. Du Pont, finally convinced of his efforts, sold him the material he needed at a reduced cost, to complete his final large piece.

A year was spent building special equipment. A special autoclave to control the curing rate of the casting was required; otherwise the casting would literally destroy itself from the heat generated during the curing process. Exact details as to how this was accomplished are not available. However, as plastic technology continues to solve the problems associated with exothermics, more castings of this size will inevitably be made. Photographs for this series by Joanne Leonard, courtesy, artist.

Bruce Beasley works on APOLYMON with a pneumatic polisher.

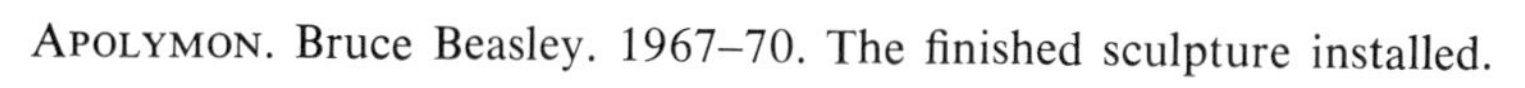

APOLYMON. Bruce Beasley. 1967–70. The finished sculpture installed.

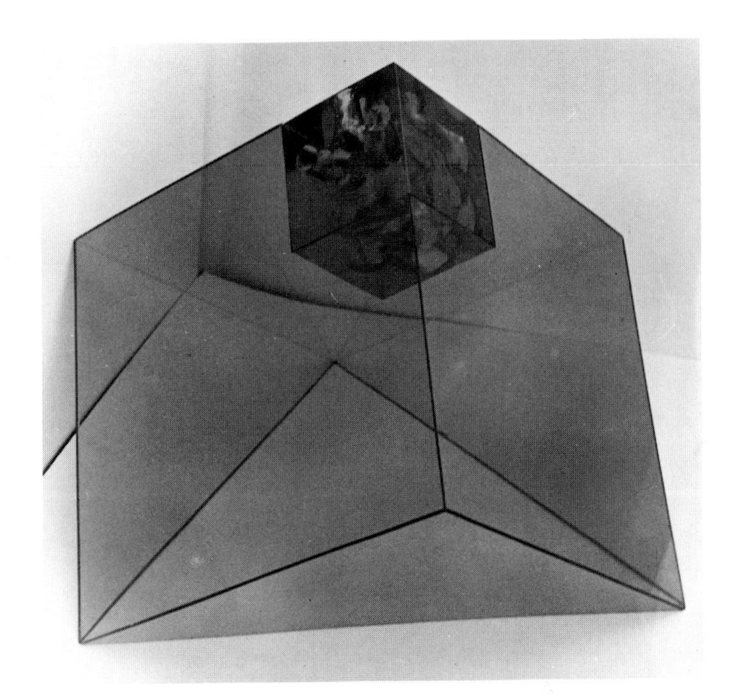

CUBED CUBE WITH NEGATIVE FORM. James Allumbaugh. 1970. Plexiglas and cast polyester, 24 inches high, 31 inches wide, 26 inches deep.
Courtesy, artist

CASTING POLYESTER VOLUMES

James Allumbaugh's sculpture is one of a series using a basic 6-inch cast polyester cube. The exterior mold for the cube was a ¼-inch plywood box coated with microcrystalline wax as a separator. The inner negative form had a ⅛-inch-thick wax wall and was attached to strategic places on the inner walls of the exterior mold to prevent its floating in the resin and to allow molten wax to drain.

Water was used to test that the mold was leakproof, then the water was blown out with a vacuum cleaner, as the smallest amount of water retards the set of polyester resin.

Resin was mixed using ⅔ clear rigid and ⅓ clear flexible polyester resins at 70°. For the 6-inch cube of approximately one gallon total, 25 drops of commercial dye and 2 drops per ounce (1 gallon equals 256 drops) of MEK (methyl ethyl ketone) catalyst were added and mixed with an electric drill.

In approximately three to six hours the exothermic heat of the gelling resin began to melt the wax which was poured out. After the casting was removed from the mold the remaining wax was removed by slowly heating the cast form in a kitchen oven at 200–250° and wiping with rags; the form was allowed to cool slowly.

An orbital sander was used, starting with a coarse paper and proceeding through stages to a 320 or so wet/dry paper to achieve flat, smooth surfaces. Hand sanding in the final stages removed any erratic patterns created by the electric sander; 320–600 grit papers were used in opposing directions. Commercial abrasive cloths start where

Micro Flat. Dennis Kowal. 1965. Epoxy, fiberglass, and ceramic glaze for color, 24 inches high, 24 inches wide.

Untitled. Peter Alexander. 1968. Cast polyester resin.
Collection, Walter Art Center, Minneapolis

600 paper leaves off and any remaining scratches are almost invisible to the naked eye. A coat of polish imparted a glossy appearance to the last surface. Commercial plastic polishes and abrasive can be used on a thick stitched cotton buffing wheel. Du Pont's No. 7 compound for automobile finishes is applicable when used with a buffing wheel at a slow speed and light pressure.

Integration with a base unit can be achieved in one of two ways: (1) precisely cut acrylic sheet and (2) plastic laminated over a plywood structure. Edges in both cases were mitered. All angles and dimensions were calculated using solid geometry on graph paper. Cast shadows and color were considered an important part of the transparent acrylic sheet bases.

TYNEMOUTH PIECE. Robert Butterfield. Cast structural polyester resin and fiberglass, 41 inches high, 40 inches wide, 22 inches deep.

Courtesy, artist

Robert Butterfield used a combination of mold techniques to create his two pieces here and on pg. 77. Direct molds were built of thin plywood and Formica with wooden members for backing to hold a particular shape. *Tynemouth Piece* (page 77) was built from one-half (one end and one side) of the base with fluted detail. Two fiberglass halves were pulled from the same plywood mold and bonded together on the edges with resin and fiberglass backing. The three top forms were built in separate molds (again using one mold for the two end pieces of lighter value). A mold was made of half the center piece, two molds pulled and joined. The corrugations were formed by pouring hot mold material (Vynamold) on a plaster cast taken from standard fiberglass corrugations which are readily available. The soft mold was warped and formed into the desired configuration.

Here Lies the Oracle was constructed in a similar manner. Mr. Butterfield used plastic corrugations and worked them over a wooden form. After the wax and release agent (liquid PVA) had been applied to the corrugations, he proceeded with the lamination process described below. A 4-inch PVC pipe was used to create the radius on the edge. After completing one side, this was used to make the opposite side. The bump shapes and ends were created on separate molds joined at the edges.

A pigmented gel coat would not produce the desired surface, so acrylic lacquer was painted on instead.

The schedule for lamination involved:

Two thin layers of gel coat with Cab-O-Sil, which allows the viscosity to be varied
One layer of glass tissue
One layer of 1½-ounce glass mat
One layer of 2- to 2½-ounce glass mat

HERE LIES THE ORACLE. Robert Butterfield. Cast structural resin and fiberglass, 84 inches high, 31 inches wide, 16 inches deep.

Courtesy, artist

This generally is enough for most structures, though larger pieces may require an additional layer of heavier mat.

Mr. Butterfield also points out that after using several types of polyester resin, he obtained the best results with structural lay-up resin manufactured by Reichhold Chemical, purchased in 55-gallon drums to reduce the cost per gallon.

Percentages of catalyst recommended by the manufacturer are best to follow, generally using the minimum percentage recommended.

Fiberglass and polyester resin tend to be unstable if not allowed to cure before surface treatment. If you plan to paint, it is wise to oven cure the piece, if possible, at 140° at least 6 hours. If oven curing is impossible, wait several weeks before painting, otherwise paint tends to dry faster than the piece underneath. Thus imperfections can occur several months after completion if a curing time is omitted. Cautions Mr. Butterfield: "Much of plastic technology has to be discovered by doing. Everyone has his own way of working."

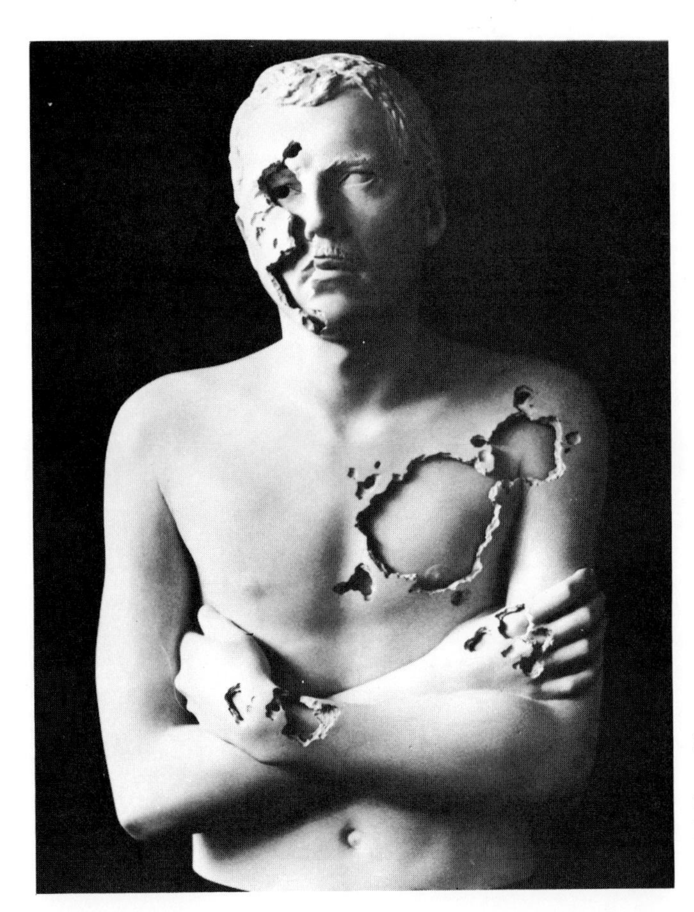

SELF-PORTRAIT. Arthur Kern. 1968. Epoxy cast by lost-wax method in plaster, 27 inches high.
Ruth White Gallery, New York

MEM. Arline Wingate. 1969. Epoxy, 2½ feet high.

Photo, Maria Martel

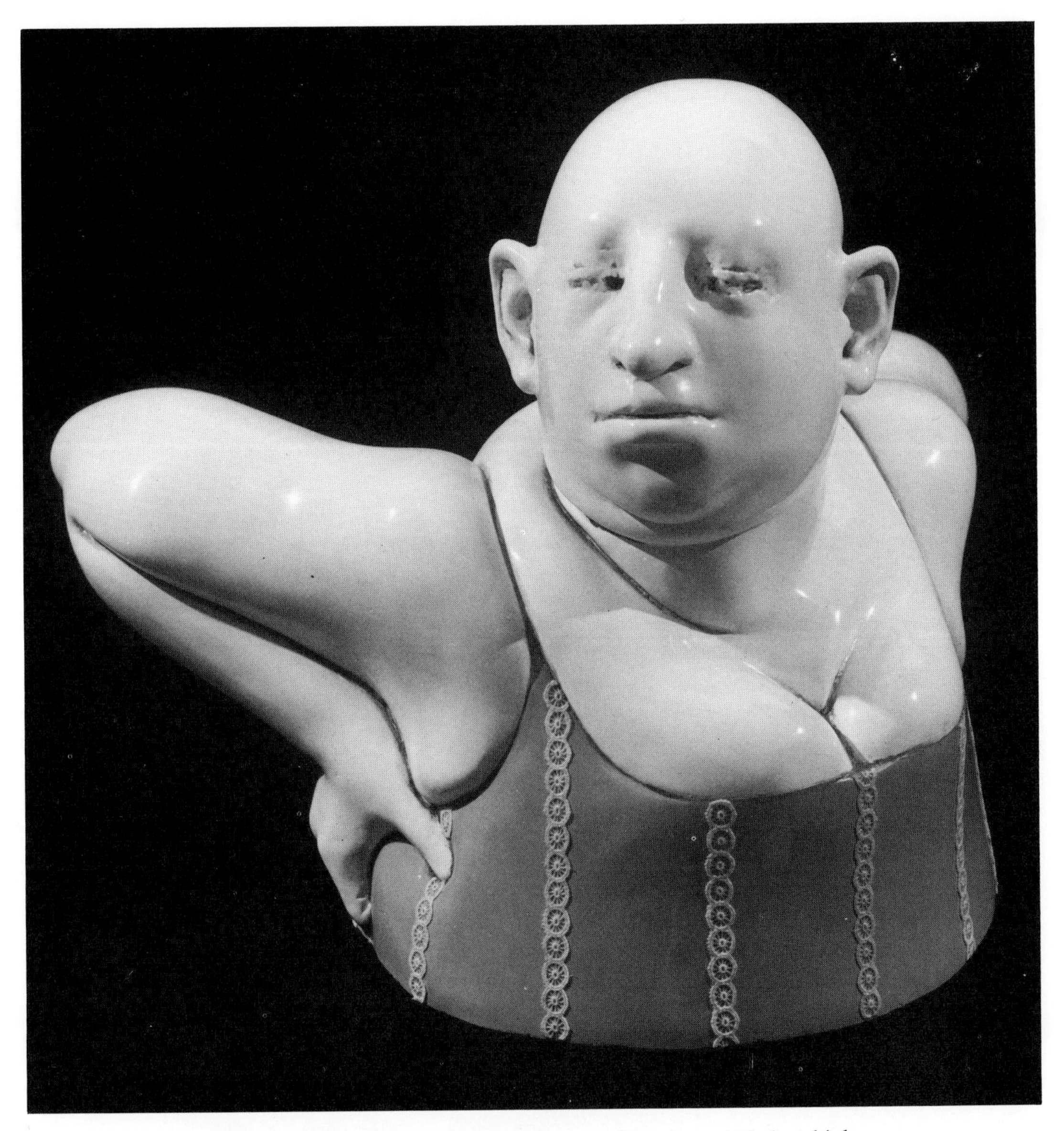

PRUDENCE. Susan Smyly. 1970. Polyester resin and fiberglass, 1½ feet high.
Courtesy, artist

A silicone rubber mold was taken directly from a plaster original. (Since silicone is self-releasing, the polyester resin was painted directly into the mold and reinforced with fiberglass.) After the casting was removed from the mold, its surface was cleaned with alcohol and sanded with 400 wet/dry paper. Two primer coats were sprayed on and sanded extensively with 400 wet/dry paper. Three coats of acrylic enamel or acrylic lacquer were sprayed on followed by hand rubbing with auto rubbing compound until the desired finish was achieved.

IDEA PER UN MONUMENTO. Giacomo Manzù. Bronze, 23¾ inches by 17¾ inches by 14¾ inches.

Courtesy, Dominion Gallery, Montreal

6

Lost-Wax Casting

THE lost-wax process of casting metal, also called "cire perdue," is probably the oldest and most traditional method of making molds. For metal casting, a rigid refractory mold is required as opposed to the flexible mold techniques described in the preceding chapters. This casting process is most often used to duplicate a sculptor's wax pattern in nonferrous metals, and excellent surface detail is one of its many virtues.

The procedure can be broken down into the following steps.

Original or Pattern

If the original pattern is of clay, plaster, or material other than wax, one of the various flexible mold materials may be used to arrive at a wax duplicate casting of the original, or a plaster piece mold is made from the original which is then used to cast a wax piece.

Once a wax pattern is made, either by casting or directly working with the wax, the sprue system is designed for the piece and attached (*see* later in this chapter).

Investment

A container of metal, plastic, or roofing paper is made and the various investment materials are mixed dry; then they are added to water. The pattern is precoated with the creamy investment and submerged in investment in the container; only the pouring cup and vents show through at the surface of the mold. Wire mesh is usually used for reinforcement, especially in larger molds and reliefs.

Burnout

Upon hardening, the investment mold is placed in a kiln and heated slowly to approximately 1,000°, which melts and burns out the wax and drives off all the moisture: this is called calcining the mold. When finished, the mold will have a cavity where the wax once was.

Casting

The mold is taken from the kiln at approximately 400° to 500° and packed immediately in sand. The metal is usually being melted concurrently and should be poured as quickly as possible. After pouring, the metal is allowed to cool, then the investment is removed and the spruing cut away. The metal—bronze, brass, aluminum, lead, or pewter—is then worked according to the sculptor's intent.

For casting, a foundry is required. A simple foundry can be arranged in a sculptor's studio or out-of-doors as illustrated in chapter 12. For those sculptors who do not have a foundry, facilities are available commercially and it is sometimes possible to use the facilities in an art school. At one time, almost all foundry work was done commercially from the sculptor's clay, wax, or plaster model. But as sculptors prefer to become involved with all processes of their art, more and more prefer to work or, at least, supervise the foundry procedure so they have more control over the result.

The following illustrations demonstrate the lost-wax casting process for small and monumental sculptures.

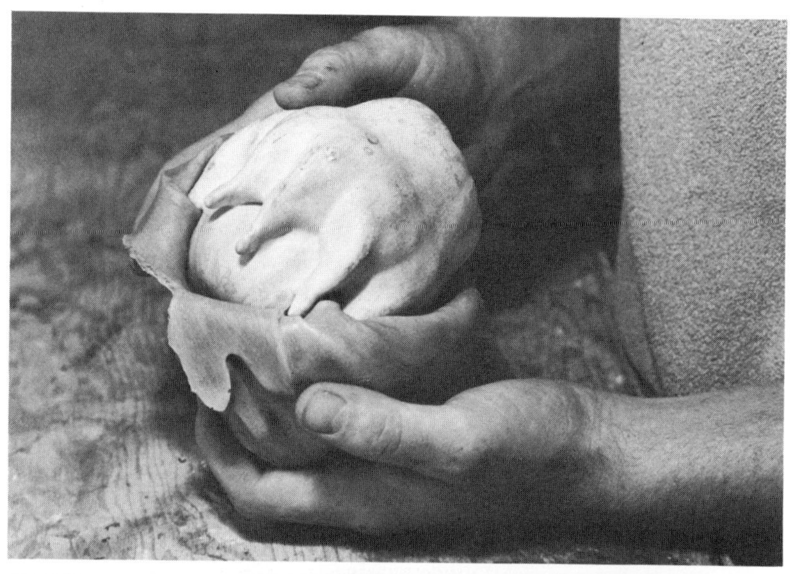

A Perma-Flex mold by Peter Fagan (see series on CMC molds) that was brushed on sets in approximately 5 minutes after the parts are mixed and a simple two-piece plaster support mold is made.

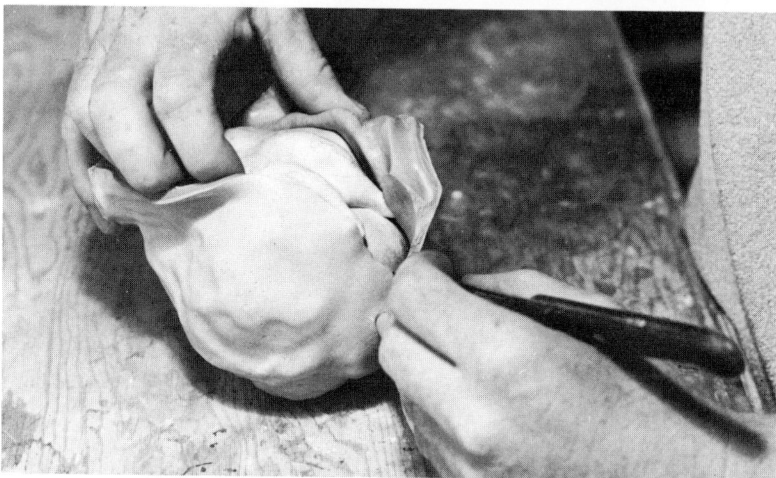

The mold is being cut to form a seam. It is peeled off as is.

The Perma-Flex flexible mold and plaster support mold with model.

Wax is melted in a pot over a hot plate to about 110°F and the temperature checked with a metal meat thermometer. Usually a microcrystalline wax is used because it works well and is relatively inexpensive.

The mold is reassembled and held together with a band cut from an inner tube. Molten microcrystalline wax is poured into the mold, sloshed around, then poured out and allowed to cool.

Repeat wax applications as many times as necessary to get the desired thickness of wax.

When the wax is cooled and removed from the mold, the thickness can be checked by holding it up to a light. Thin areas can then be thickened. The wax should be approximately 1/8 to 3/16 inch thick.

To retouch wax, use an alcohol lamp and a metal tool.

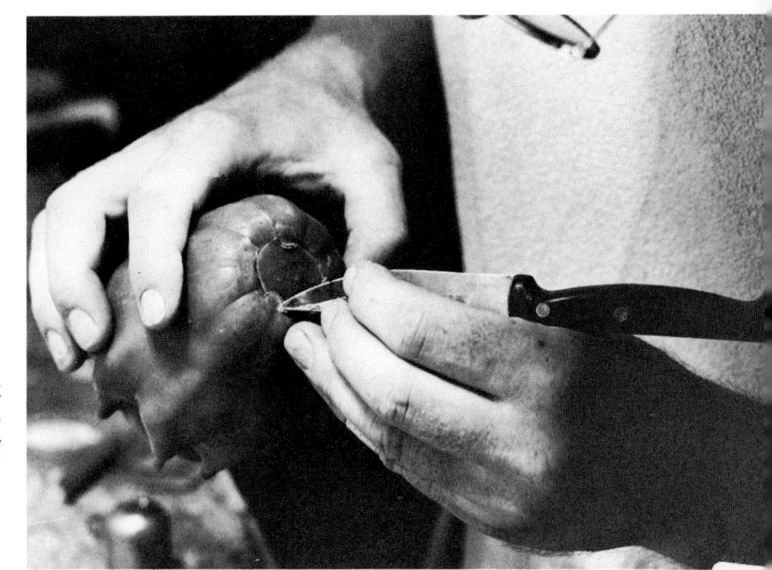

The plug that will be cast separately is cut and fitted to an opening in the bottom of the piece.

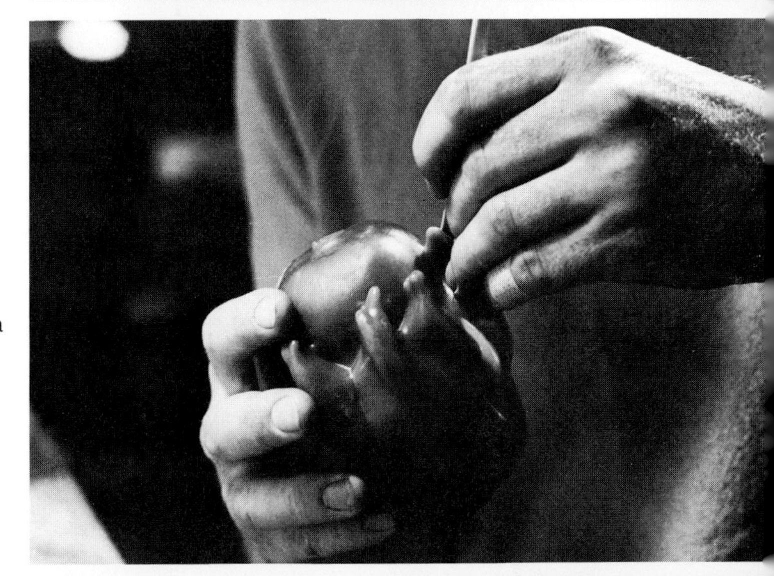

Wax is reworked with a wooden tool.

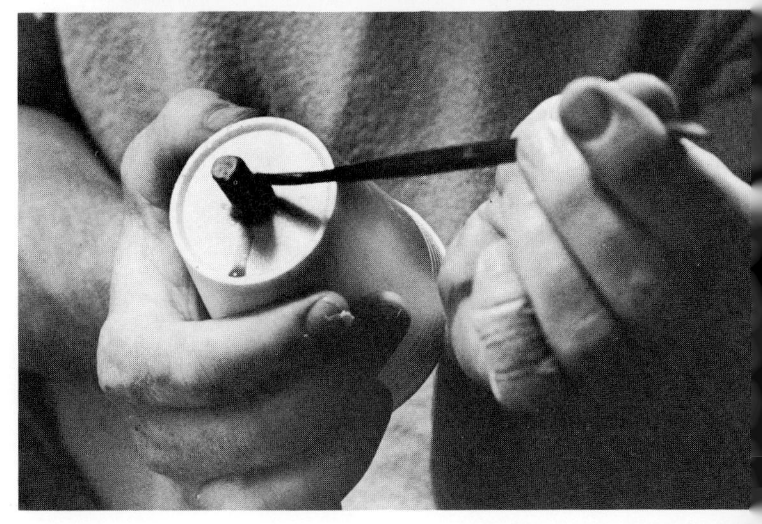

The sprue is attached to the pouring cup.

INVESTMENT CASTING

Investment casting is the process by which an original piece, generally in wax but sometimes foam, natural fiber, paper, or light wood such as balsa, is incased in a binding mixture such as plaster with a refractory material, silica sand. The mixture is allowed to harden and then baked in a kiln to melt or burn out the original and dehydrate the investment. Through openings that connect the piece to the surface, metal is poured into the investment mold and gases escape.

A good mold will retain the surface quality of the original and resist washout from hot metal flowing into the mold. A standard basic mixture of plaster (binder) and sand (refractory) can be varied according to availability and additives such as crushed firebrick or old investment, dry fireclay, grog, and Perlite. The size of the casting, weight, cost, burnout facilities, and the metal poured are considerations as to investment materials and proportions to use. Most investment material is mixed dry and by hand; for large amounts a portable cement mixer can be used. The dry mix is then added to water.

The investment can be poured around the wax placed within a flask or container of metal, brushed or laid up by hand or with hand tools. Chicken wire, hardware cloth, or metal rods can be used for reinforcement. Reliefs demand special attention as they are thin and tend to collapse.

A wax original with sprues attached and ready for burnout, 35 inches long. A foam pouring cup is at left. George Greenamyer.

Courtesy, artist

Investment is mixed in a wide flexible pail using equal parts silica sand and plaster by volume. Mix ingredients dry, then sift evenly over surface of water until mounds build up on surface; let stand for a few minutes and mix by hand or with an electric drill attachment.

UNTITLED. David Hendricks. Epoxy and leather.

UNTITLED. Steve Daly. 1971. Cast bronze, aluminum, and iron.

SWAN AND ITS WAKE I. Jack Zajac. 1968. Bronze.

Students at Hinckley-Haystack School of Arts and Crafts pour aluminum into a sand mold.

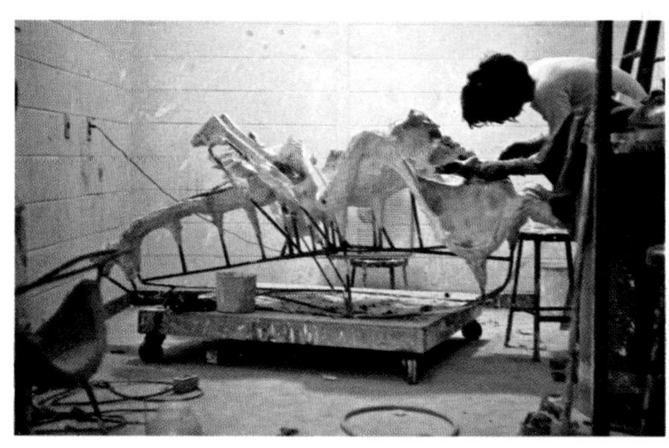

LIFE-SIZE PLASTER WASTE MOLD. David Hendricks.

BISHOP PANEL. Tom Tasch. Epoxy.

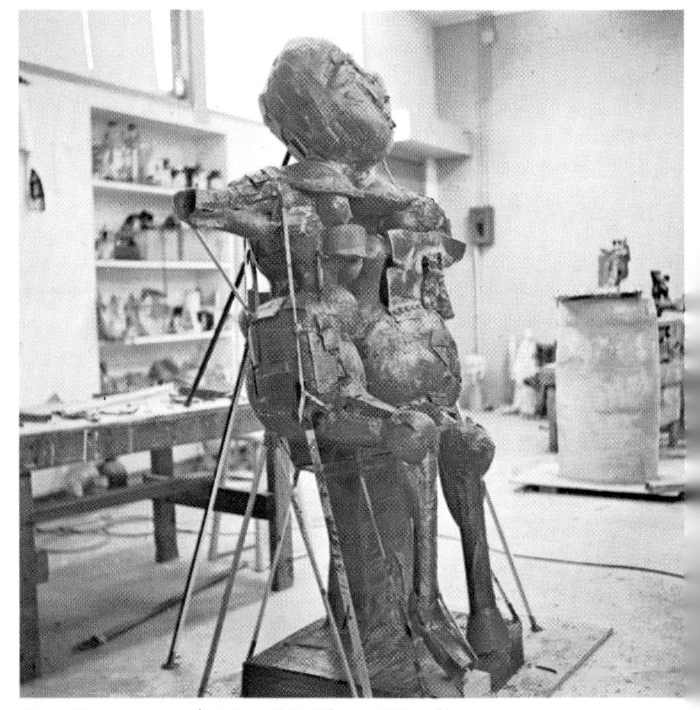

EVE (in progress). Tom McClure. Wax figure.

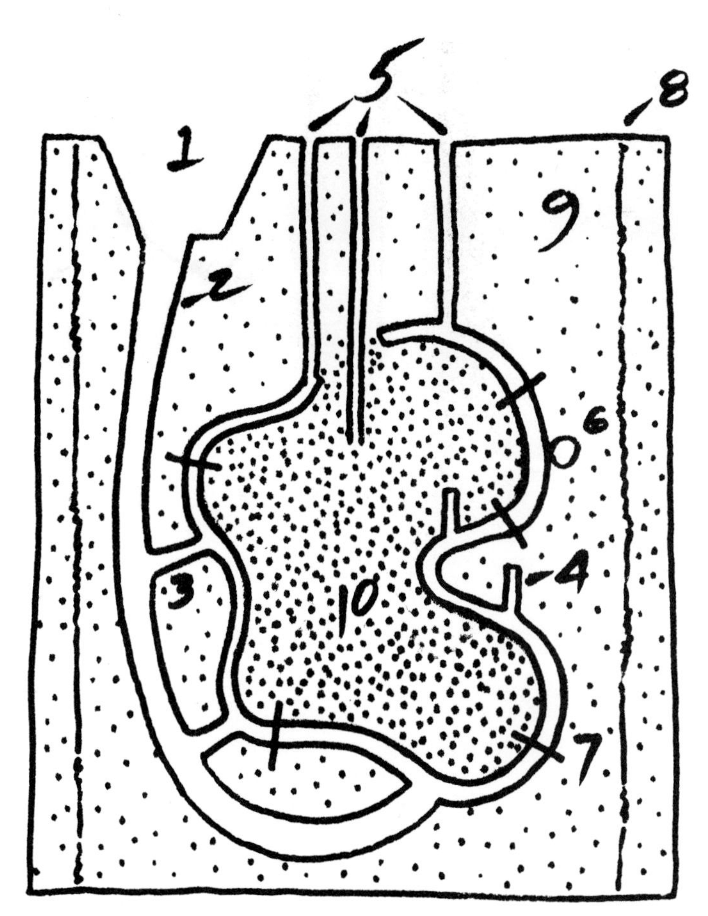

Drawing, Lebbeus Woods

1. Pouring cup—receives metal poured from crucible.
2. Sprue (main)—carries metal from pouring cup to piece. Where it is attached to the piece it is called the "gate."
3. Runner—used to carry metal to predetermined levels of piece.
4. Riser—acts as a reservoir to absorb reverse flow of metal as it cools and shrinks. It can also act as vent located on thick areas; can be called "live riser."
5. Vent—allows gases to escape and indicates when mold is filled and can act as a riser. It is located at high points or where extra flow of metal is desired; also "vent core."
6. Spherical blind riser—largest metal supply with minimum surface contact.
7. Core pins—usually of same metal, used to hold core in place when wax original is melted out and metal is poured into mold.
8. Wire or other metal reinforcement impregnated in investment.
9. Investment—refractory material and binder; sand and plaster, for example.
10. Core—same material as on the outside investment.

Note: Pouring cup, sprues, vents, etc., can be made of wax with a lower melting point than original to facilitate meltout.

Typical Investment Materials

grog (ground fired clay) *
luto (used investment) *
molding plaster
brick dust *
clay *
silica sand *

---

* Refractory material.

Relief shown immediately after devesting prior to chasing. By Peter Fagan.

An investment core is poured; wire will be inserted into the core before the investment material sets.

Zona-lite (mica) expanded volcanic ash vermiculite and Perlite (used to lighten mold)

Typical Proportions

1 part plaster—1 part silica sand

1 part plaster—1 part silica sand—1 part luto

1 part plaster—1 part fireclay—1 part grog

1 part plaster—1 part fireclay—1 part Perlite

⅓ part plaster—⅔ part silica sand

¼ part plaster—¾ part fireclay

1 part plaster—2 parts river sand

A surface coat mixture such as 1 part plaster, 1 part silica sand, 1 part fireclay can be applied first to assure fine surface detail, then followed by a cheaper mixture for strength and weight.

A Typical Sprue System Used in Investment Casting

Core pins may be placed in piece with only a point touching investment through the wax, leaving a fine point if anything to chase off after casting.

The wax with core poured and pouring cup attached. The plug that will fill the hole is attached to sprue. The wire will help suspend the piece while it is being invested. A foam cup is used for the pouring cup.

Core pins of steel or brass are added.

A wax piece with its underside invested showing wooden sides (which will burn out along with the wax) with brass nails for core pins.

A simple-to-make plaster mold is used to form sprues and vents used for investing. Wax is poured into wet molds to prevent it from sticking. Similar molds are used to make forms or shapes for sculpture.

Galvanized metal sheeting is used as a container for the investment; chicken wire reinforcement is fitted to the inside leaving room for wax piece. A small amount of investment is poured into the bottom of the container to secure container and reinforcement.

Investment is poured into the sheet metal container. The wax form is coated and checked for complete surface coverage.

The piece is gently pushed down into the investment and held there until the investment sets, approximately 5 minutes.

After the investment sets, the metal is stripped from the mold.

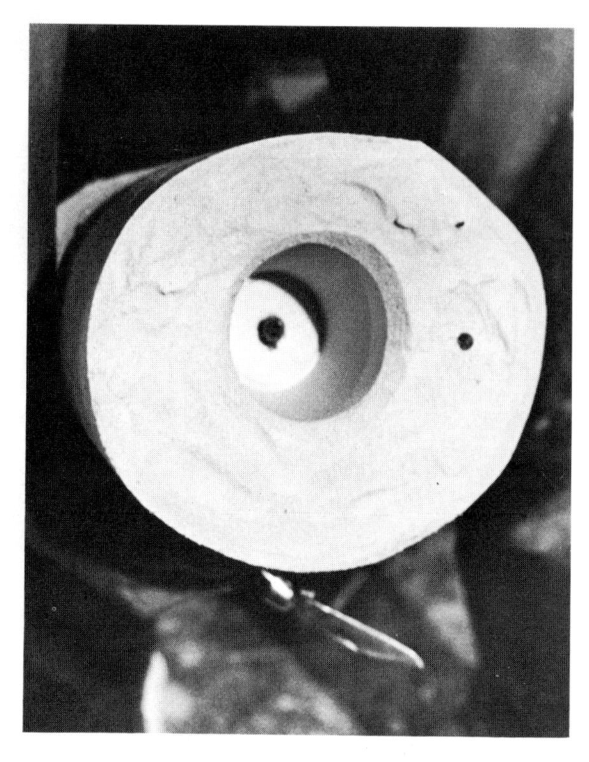

The pouring cup is cleaned out and the vent scraped clean.

Mold is packed in a burnout kiln. A roll-out door with a platform attached to it facilitates loading and unloading. The burnout is controlled by a wall-mounted thermostat.

Another view of the electric burnout kiln showing the rolling door with platform and the molds loaded. The kiln is slowly heated to 1,000°, held there for one day, and cooled down slowly—three days in total.

Mold after wax has been burned out. A thermo-crack in the mold shows up on the casting as a fin or flashing.

Molds are placed in a pit with sand and the mold opening covered with aluminum foil to prevent foreign matter from getting in. Sand contains and stabilizes the mold during pouring. If wax or moisture is still in mold when metal is poured, an explosion could result. Packing in sand prevents this but the metal could still shoot up: A good reason for wearing protective headgear when pouring.

Ingots of bronze are placed loosely in the crucible before the furnace is fired; the crucible rests on a refractory block. *Do not* jam metal into crucible; as metal is heated it expands and could break the crucible.

GENERAL PROCEDURE FOR MELTING AND POURING METAL

1. The crucible * is placed on a silicon carbide crucible rest inside furnace.
2. Ingots or scrap metal are placed inside the crucible. *Never* force or jam metal into the crucible, as it expands during heating and could crack the crucible.
3. Start furnace and adjust; lower top of furnace and hood, if one exists, over furnace.
4. All metal or tools to be added to molten pot must be *preheated* to drive off all moisture, which could cause an explosion. Place additional metal on top of the furnace or in a preheating chamber, then lower it into the crucible with tongs to avoid splashing metal, which is dangerous and can ruin furnace walls.
5. Never stir molten metal; the less it is disturbed the better, except for skimming. Avoid reducing flames (too little air).
6. The best way to check the temperature of the metal is with a thermocouple connected to a pyrometer which measures the temperature of molten metal below its surface.
7. The metal may be skimmed while still in the furnace by reducing flame, skimming and reheating, especially if scrap is used. Do not hold metal at pouring temperature or overheat.
8. When metal is at desired pouring temperature, flux metal if necessary. Shut down furnace, raise and remove lid, remove crucible with crucible tongs. Place in crucible pouring shank on refractory fireblock, skim, then pour metal.
9. Pour excess metal into ingot molds for future use.

* American standard shape graphite crucibles have an approximate capacity of one pound of aluminum or three pounds of copper-base alloy to the number. Therefore, a number 100 crucible would have a working capacity of 100 pounds of aluminum or 300 pounds of copper-base alloy. To determine its capacity for other metals, multiply the water capacity by the specific gravity of the desired metal. Deduct 10 percent to obtain working capacity.

The furnace is gas fired.

Scrap metal is preheated before lowering into crucible.

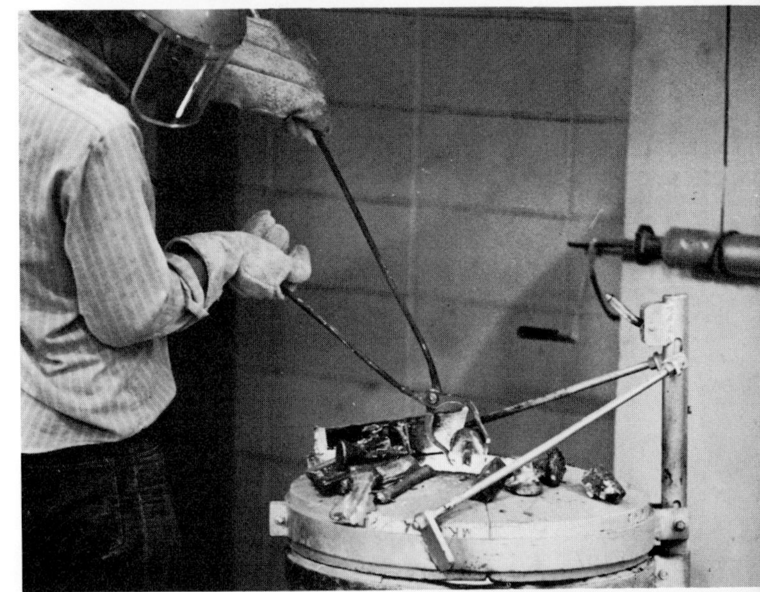

Temperature is taken with a wall-mounted pyrometer. Thermocouple is connected to pyrometer by cable.

After furnace has been shut down, a two-man crucible tong is used to remove crucible. Note that the metal was not fluxed as silicone bronze was used; however, skimming was required.

Metal is skimmed (to remove foreign material, slag, and dross from surface of molten metal). The crucible rests on a refractory material in a two-man crucible shank with safety latches.

Foil is removed from holes of mold immediately before pouring. As metal is poured, one man controls the pour, the other follows.

Pour excess metal into prewarmed ingot molds. Metal in ingot form is easy to store and clean.

After metal cools, small molds are removed from sand with tongs; larger molds require an overhead lift.

After cooling, investment is broken off to release metal form. Here it is partially broken to show reinforcement and pouring cup.

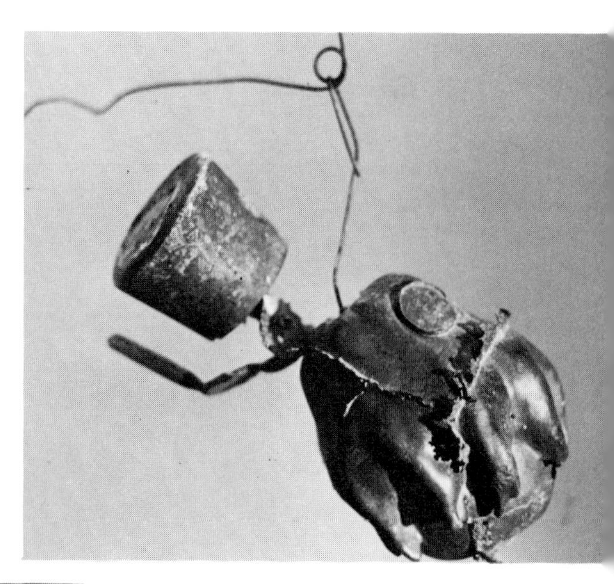

Bronze casting after investment has been knocked off. When removing investment, be careful not to hit metal. Pouring cup, vent, and flashings will be cut off, the patch fitted into the hole after the core is knocked out, and holes from core pins filled or hammered shut. If desired, a patina may be applied, but often the metal itself can be beautiful just as it comes from the mold. The type of wax used, the temperature at which the metal was poured, whether the mold was completely burned out, the type of alloy used, all have an effect on the color of the metal.

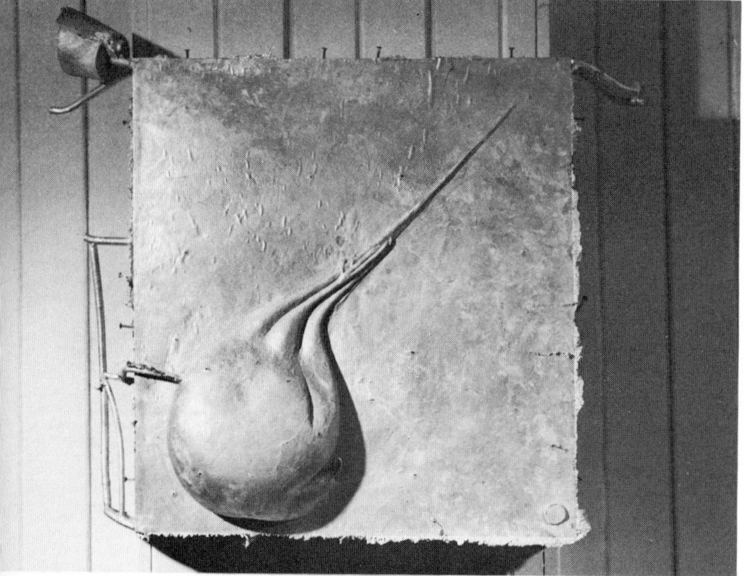

Detail from photograph on page 91 after it was cast in aluminum. Observe core pins and location of pouring system.

A completely portable furnace made by Peter Fagan. Firebrick is cemented in place with refractory cement inside a steel jacket. Lid is mounted on a hinge and can be swung to the side.

A used deep-fat fryer may be adapted for melting wax.

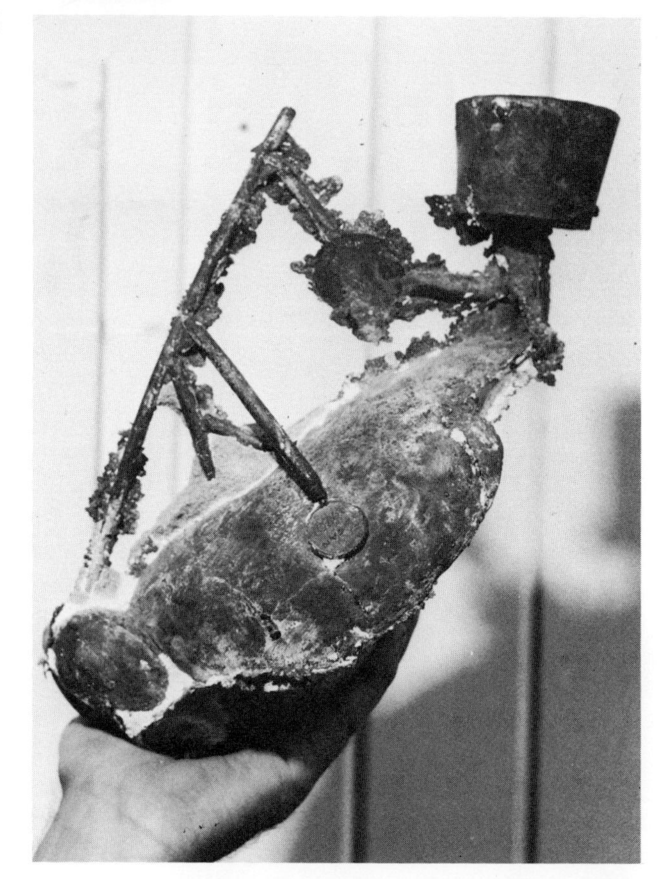

Detail of a bronze that has been removed from its investment with sprue system still attached.

Close-up of a blower converted from a vacuum cleaner motor used for air supply on the furnace shown on page 98. A swinging valve (front) controls the air flow. Further up the black pipe a valve has been fitted to control flow of gas. This unit can be fired with natural gas, propane, or oil.

THINGAMAJIG. Richard Herr. Discarded machined auto parts are combined with a modeled section. This sculpture was created by modeling clay over a discarded auto part. A plaster mold was made over both pieces and the modeled section removed. The mold was then calcined at 450°F for 12 hours and the aluminum poured which adhered to the auto part. A refractory material such as silica sand is generally used for larger aluminum castings.

► Plaster for bronze casting. William Fothergill. 1971. Over life size.

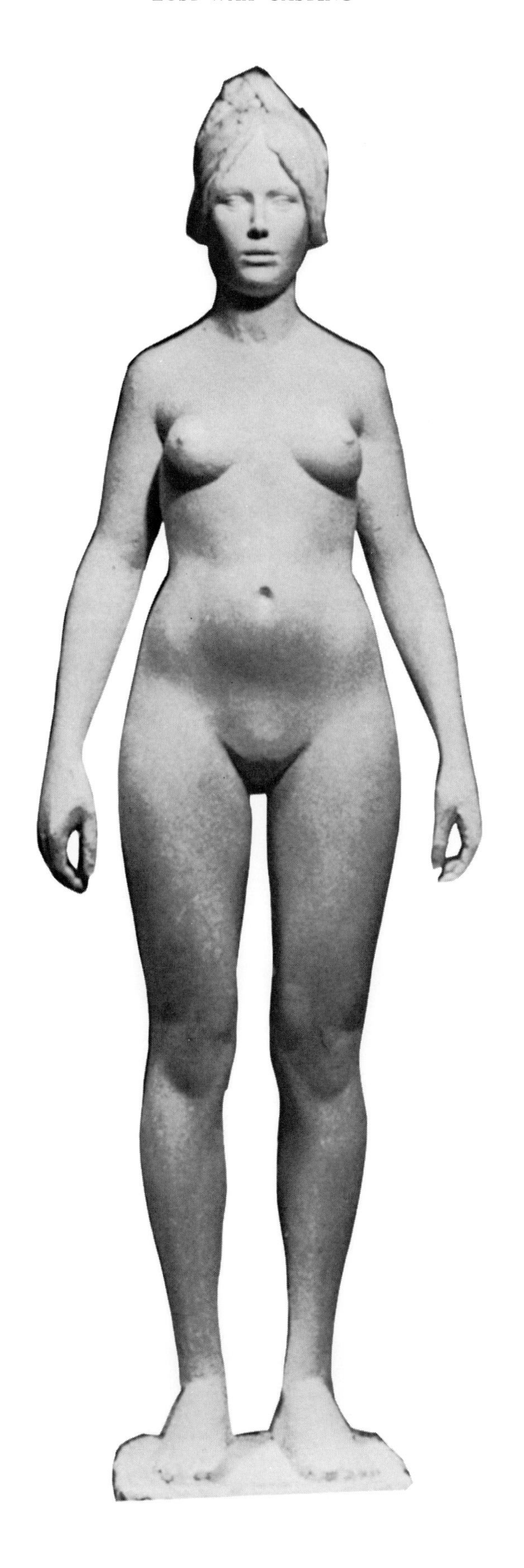

FRAGMENTED PORTRAIT. Dennis Kowal. Bronze, approximately 18-inch diameter.

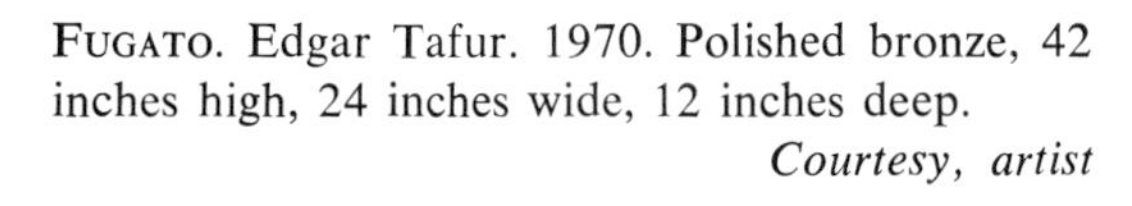

FUGATO. Edgar Tafur. 1970. Polished bronze, 42 inches high, 24 inches wide, 12 inches deep.
Courtesy, artist

CROW-TERM. Leonard Baskin. 1968. Bronze, 36 inches high, 6½ inches wide, 10½ inches deep.
Courtesy, Grace Borgenicht Gallery, New York

WIND AND WAVE. Thea Tewi. 1962. Bronze, 21½ inches high, 20 inches wide, 10 inches deep.
Photo, Jack Lessinger

KNIEENDE (KNEELING FIGURE). Wotruba. 1960. Bronze, 63 inches high.
Courtesy, Marlborough-Gerson Gallery, New York

WINTER BRIDE. Elbert Weinberg. 1965. Bronze, 53 inches high.
Courtesy, Grace Borgenicht Gallery, New York

LIFE. Virginio Ferrari. 1969. Bronze, 22 feet high.
Courtesy, Loyola University Medical Center, Hines, Ill.

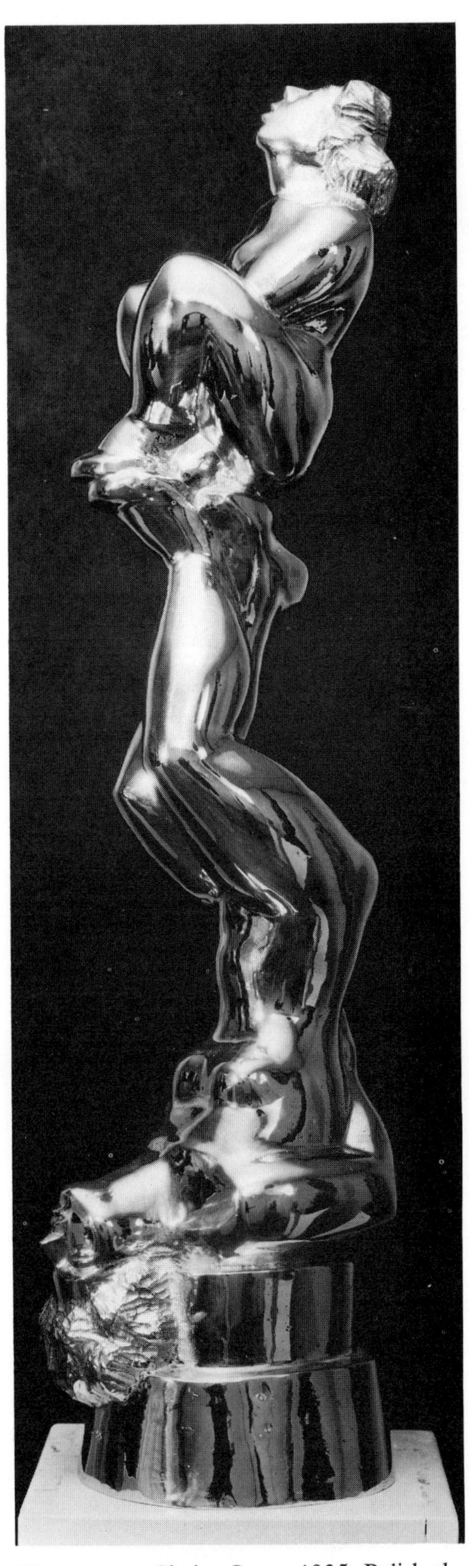

BALANCING. Chaim Gross. 1935. Polished bronze, 34 inches high.
Collection, Whitney Museum of American Art, New York

UNTITLED #2. David Slivka. 1961. Bronze, 11½ inches high, 16 inches wide, 10 inches deep.
Collection, Walker Art Center, Minneapolis

READING ARM. Elbert Weinberg. 1969. Bronze, 28½ inches high, 44 inches wide, 15 inches deep.
Courtesy, Grace Borgenicht Gallery, New York

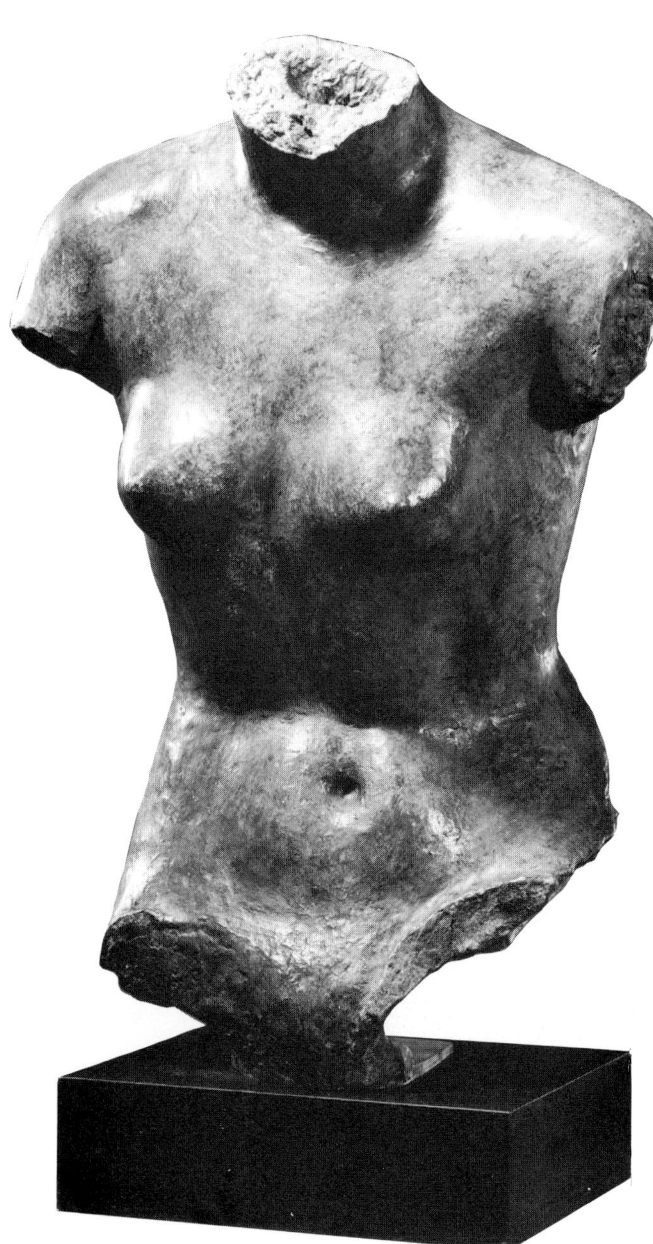

TORSE ANTIQUE. Aristide Maillol. 1902. Bronze, 27 inches high.
Collection: Walker Art Center, Minneapolis

MOSES. Sorel Etrog. Bronze.
Photo, Dona Meilach

DANCING WOMAN. Paul Zakoian. 1963. Bronze, 26 inches high.

Courtesy, artist

WALL. Helen Beling. Bronze.

Photo, Walter Rosenblum

HELIXIKOS #3. Hans Hokanson. 1969. Bronze from a wood carving, 39½ inches high, 24 inches wide, 25 inches deep.

Courtesy, Grace Borgenicht Gallery, New York

TALISMAN. Jason Seley. 1969. Bronze, approximately 12 feet high. Installed Casper College, Casper, Wyoming, 1969. Cast at Modern Art Foundry.

This piece was done originally in welded metal. Flexible molds were made to cast a wax piece, at which time areas thought inappropriate for bronze were reworked in the wax and the bronze made by the lost-wax method. After casting, the piece was cleaned to the raw metal and left out-of-doors to develop its own patina as the winds and sand would have had an undesirable effect on an applied patina or finish.

Courtesy, artist

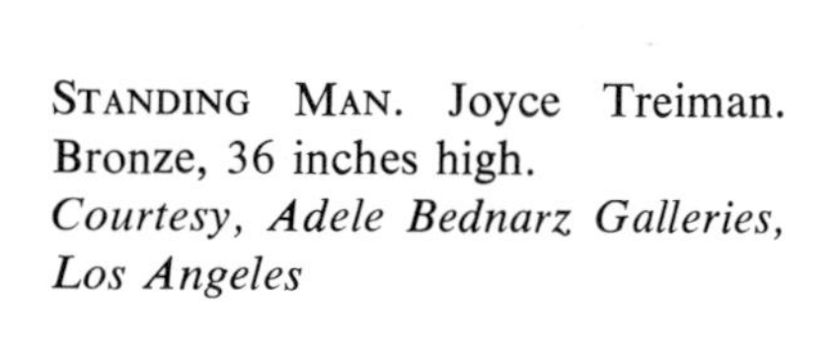

STANDING MAN. Joyce Treiman. Bronze, 36 inches high.
Courtesy, Adele Bednarz Galleries, Los Angeles

PROMETHEUS STRANGLING THE VULTURE II. Jacques Lipchitz. 1944–53. Bronze, 100 inches high.
Collection, Walker Art Center, Minneapolis

BIG SKULL AND HORN IN TWO PARTS. Jack Zajac. 1963–64. Bronze, 72 inches long, 16½ inches wide.
Photo, W. B. Nickerson

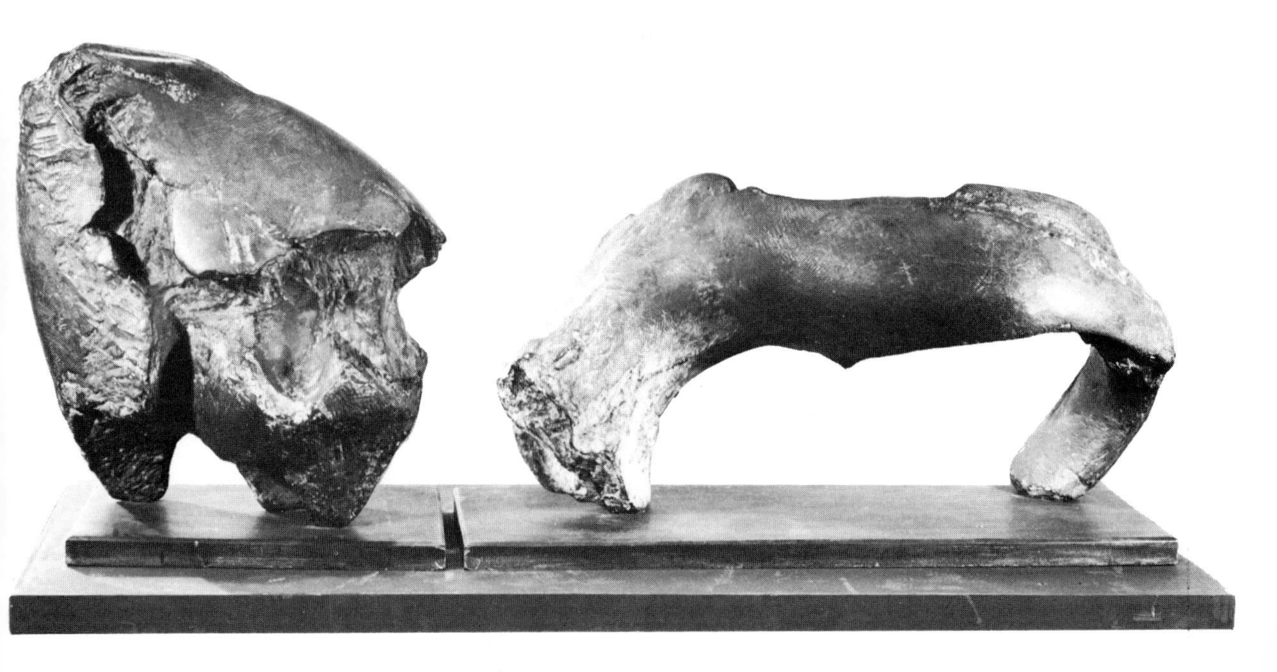

THREE PIECE RECLINING FIGURE. Henry Moore. 1960s. Bronze.

The Art Institute of Chicago

BOUND GOAT. Jack Zajac. 1957. Bronze, 28 inches high.

Photo, W. B. Nickerson

WOMAN TORN. Elizabeth Model. 1965. Bronze, modeled in wax.
Photo, O. E. Nelson

VIRGINIA WOOLF WITH ASPHODEL. Leonda Finke. Bronze, 11 inches high, 8 inches wide.
Courtesy, Harbor Gallery, New York

ASCENDING ANGEL. Katherine Nash. 1965. Bronze, 15 inches high.
Collection, Mrs. Margery Howard, Excelsior, Minn.

Left: Original plaster.

Below: Piece mold partially assembled with a mother mold that holds all the smaller interlocking pieces taken from the original. Wax, a mixture of beeswax, pigment, and paraffin, is poured into the wet mold; the mother mold and then the smaller pieces are removed. The wax model is invested with a mixture of 1 part plaster to 3 parts burned plaster and ground insulation bricks.

In the following series, Leo Kornbrust, Germany, demonstrates an unusual burnout furnace that is fired by wood instead of the normal fuels: natural or propane gas or oil.

Right: Kiln in operation is stoked with wood for one to two days and nights.

Below: Several molds stacked (pouring cup down) on firebricks with the first layers of brick for the burnout kiln being built up around the molds.

After burnout is completed, a mirror may be held near an opening in the bricks, and if condensation forms on the mirror, the molds are still wet. The kiln is dismantled.

The hood has been removed from the furnace. Observe that the furnace shown is a tilting model, but it was not used that way. The crucible was lifted out and placed in a two-man crucible shank. The crucible is resting on a refractory material.

The furnace is fired long before the burnout is completed so the melting of the metal corresponds with the completion of the burnout. Furnace is fired by wood inserted in the opening toward the bottom of furnace.

Following completed burnout, molds are removed from the kiln, and boards are clamped around the investments to secure them against bursting. They are lined up according to a pouring sequence and placed on sand. Note the smooth pouring operation.

After the bronze has cooled, the investment is broken off. The form is seen with the pouring cup, sprues, and vents still attached. Water is both cooling metal and washing off investment, which is not always advisable as rapid cooling anneals bronze.

Photo series courtesy, Leo Kornbrust

DIMITRI HADZI CREATES *THERMOPYLAE*

The following series of photos, covering more than a year's work by sculptor Dimitri Hadzi, shows the events involved in creating the 16-foot-tall sculpture *Thermopylae,* commissioned for the J. F. Kennedy Federal Office Building in Boston. 1969. Photos: D. Hadzi, Max Barus, Jr., Frank J. Monaco, Ronald Rosenstock.

A small bronze cast (*left of photo*) made from the original wax model is pointed up with a mechanical pantograph to the half-size 8-foot plaster model.

The 8-foot plaster model was transported to the artist's studio where it was reworked for a six-month period.

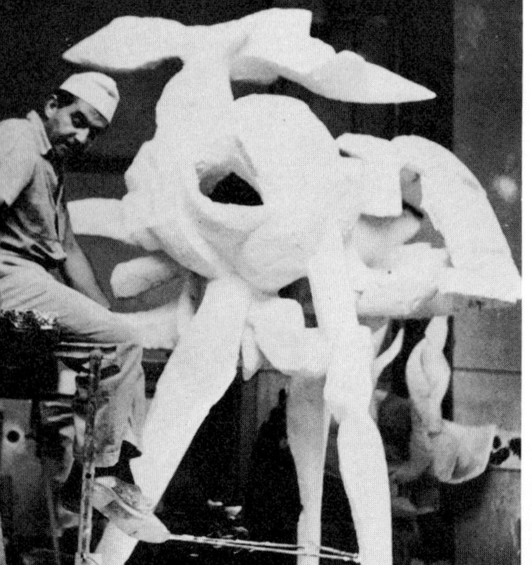

Burlap is laid over a steel armature to provide reinforcement for the plaster. The half-scale 8-foot plaster model was cut into pieces and mechanically enlarged piece by piece to the full scale, 16 feet. The craftsman is pointing up from the model a full-scale steel armature.

Another view of the enlargement showing the reinforcement necessary with such a large piece. Note the half-scale model in the foreground.

Plaster piece molds were then built up around the finished piece and in many sections. The finished piece mold sections are then removed and coated with wax on the inside to the thickness of bronze desired. Hadzi spent a full year working the plaster before the piece mold was made.

The molds, coated on the inside with wax, were assembled and the hollow space, or interior, filled with a clay-based investment casting material which is the core. The outer plaster molds were then removed and the wax was retouched.

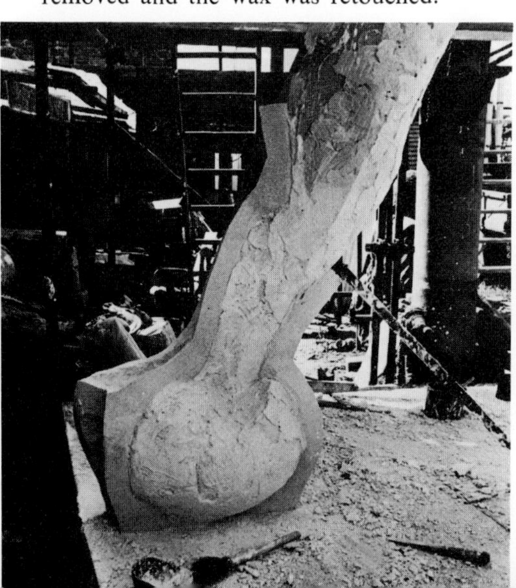

The gates and vents were then attached, nails were added to act as core pins to hold the core in proper relationship with the final outer mold after the wax is burned out.

The outer side of the wax sections are now coated with investment that is reinforced with steel wire and strapping to prevent the mold from bursting during casting. The molds are then burned out inside a subfloor burnout pit. After the molds are burned out, the metal is poured.

The investment is removed from the castings after cooling and the spruing system removed along with the nails. The holes are plugged. Individual sections are assembled and carefully welded along precise seams. Twenty sections were required for this piece. Note the "Roman" sleeve-type joint which is used in several places to permit easy, on-the-site assembly and accuracy.

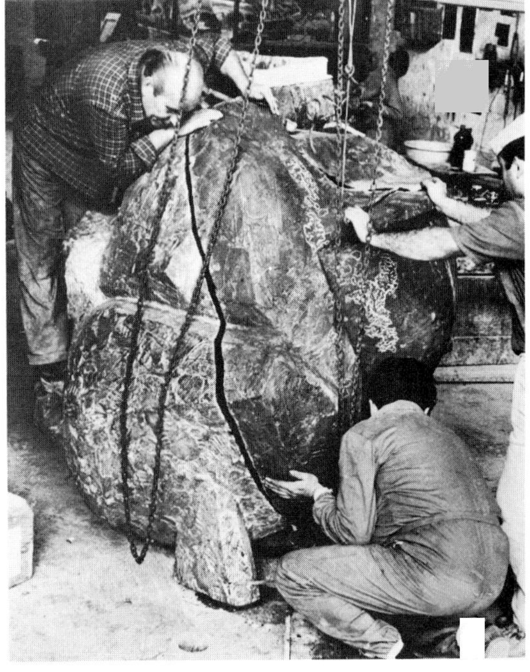

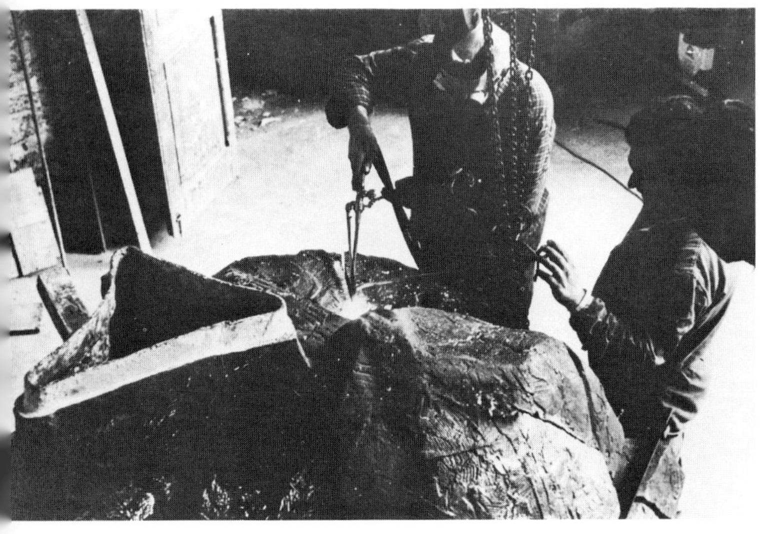

Sections are assembled at the foundry. Note holes in the Roman joint at the upper left. These holes accept special bolts and pins that hold the section securely in place.

Piece being crated at foundry.

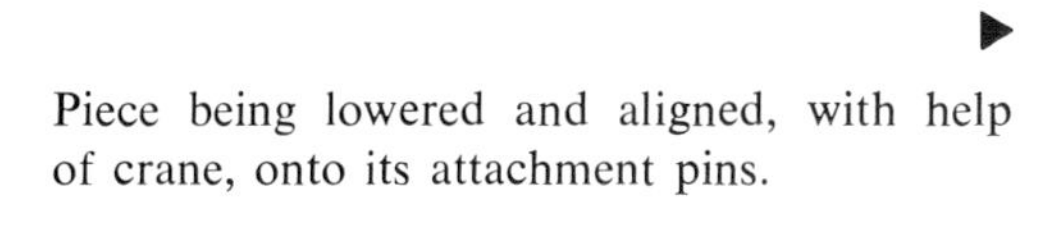

Piece being lowered and aligned, with help of crane, onto its attachment pins.

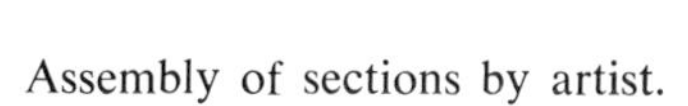

Assembly of sections by artist.

The finished piece. THERMOPYLAE by Dimitri Hadzi. Bronze, 16 feet tall.

7

The Gelatin Mold

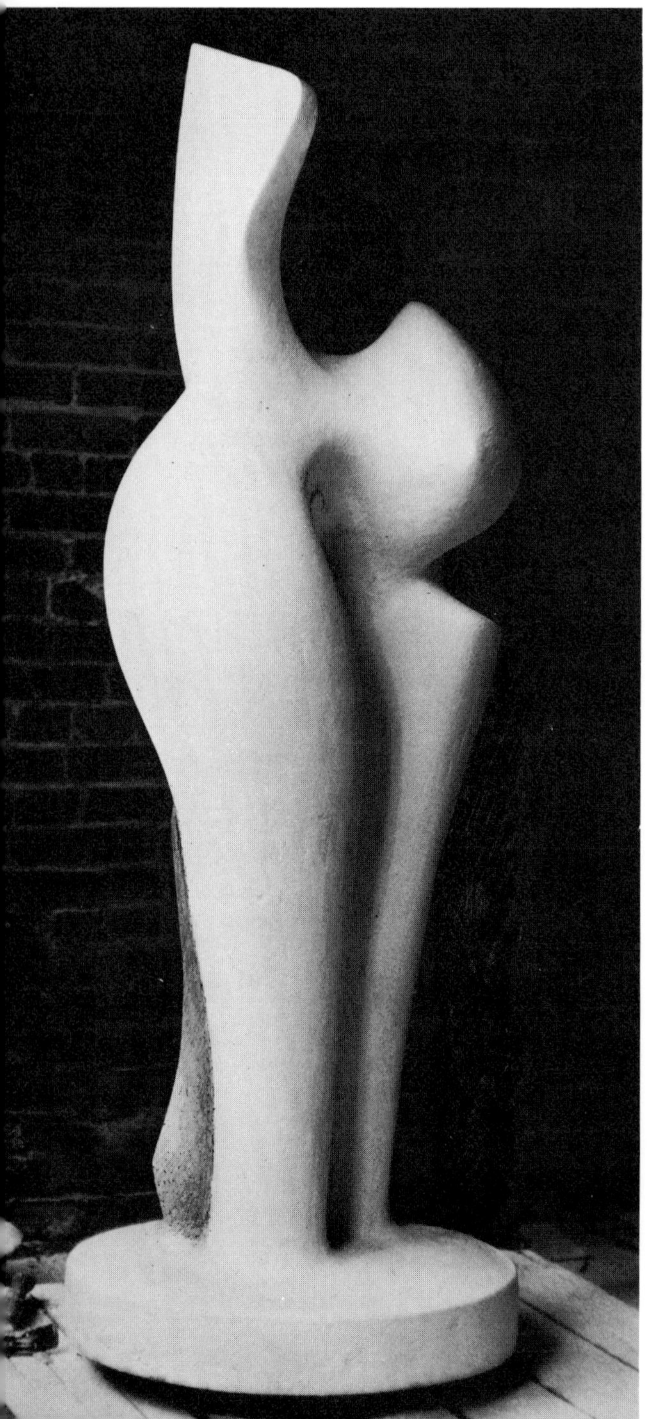

SINCE the advent of plastic and rubber molding materials, the gelatin mold is not as popular as it once was. Yet, in many situations, one will elect to use the gelatin mold because it is inexpensive compared to the other materials. At this writing, a gallon of gelatin costs approximately a third that of latex, which is approximately one-fifth the cost of *silicone rubber* per gallon. For some plaster casting, gelatin is an efficient material. The sculptor, seeking the broadest knowledge of materials available to him, should be familiar with its properties.

A gelatin mold is sometimes referred to as a glue mold, but the gelatin used for molding is a purer form than glue and possesses more elasticity.

Two kinds of gelatin can be used for molding. Agar is a vegetable gelatin secured mainly from marine algae. It is a colorless and amorphous solid that softens and liquefies when it is dissolved in hot water. It is a comparatively thin and delicate material for molding and requires a plaster support mold. The agar base material is a relatively recent development that emerged during World War I as a material for taking molds from flesh because it hardens slightly above the normal human body temperature. For sculpture, agar can be used as a negative mold material to contain plaster of paris, portland cement, and wax. Most agar-water

AURORA. Dennis Kowal. 1965–66. Original plaster, 6 feet high. Built on a welded steel armature.

mixtures require a body strengthener for thickening and preservatives to prevent attack by bacteria.

The second gelatin material is extracted by boiling and processing the skins and bones of animals; it carries the technical term collogen. Only the first cookings that yield the finest gelatins are used for making molds because a tough material results. Gelatin is generally available in sheets, flakes, and as a fairly coarse powder which is quite brittle when dry and soluble in hot water.

To prepare gelatin for use as a negative mold, it is broken into small pieces and placed in a jar with enough cold water to cover the mass and then allowed to stand for about 12 hours or overnight. The gelatin will slowly absorb water and soften. The entire contents of the jar is then placed in a double boiler and heated slowly until it is thoroughly melted. It is removed from the flame and allowed to cool to body temperature; any resulting scum can be removed with a ladle. The mixture should not be poured when it is too hot; proper pouring temperature can be determined by dipping your moistened finger into the mass; when it is a comfortable temperature, approximately 120°, it is ready for pouring.

The following series illustrates the procedures for taking the gelatin mold from a plaster original. The finished mold will be flexible; it is removed by folding back to recover the original. The inside of the mold is then washed with a strong solution of alum to partially insolubilize the surface of the glue. This is necessary because any heat generated by the mixture that will be poured into it will tend to fuse the plaster with the gelatin. After the alum wash, it is advisable to spray the inside surface of the mold with a silicone release agent.

Usually several castings can be made with a gelatin mold before the details in the reproduction start to become distorted because of the heat from a wax mix or heat and moisture from a plaster mix. For this reason, gelatin molds are rarely used by industry anymore; however, when there is little reason for using a mold many times, gelatin can often be an inexpensive, efficient solution to mold making.

Some catalogs of sculpture materials feature the product Moulage, ready-made to melt in a double boiler. It can be reused often, as additives are available to reconstitute the original purity of the molding material. Some moulage products have a wax base and others have a vinyl base, depending upon the formula of the individual manufacturer for the gelatin or moulage mold materials. All reputable products include instructions for proper use.

The following series illustrates how a gelatin mold was made from the original plaster sculpture *Aurora* by Dennis Kowal. Mold demonstration by Paul Zakoian. Photos, Cal Kowal at the Contemporary Art Workshop, Chicago.

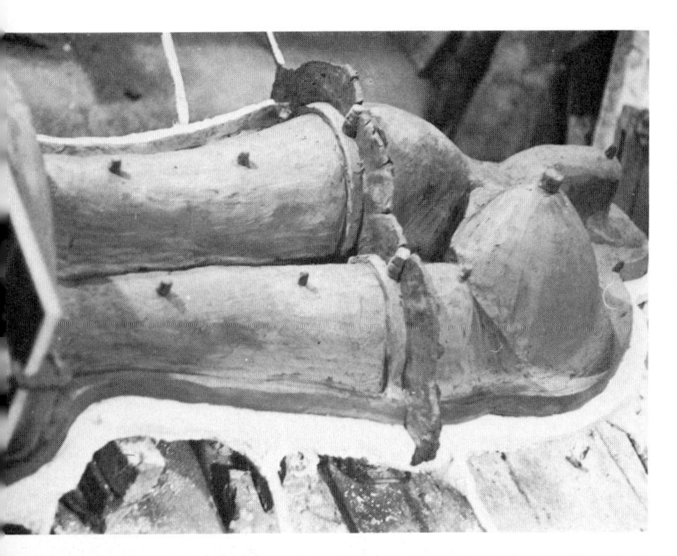

The original plaster piece, which has been sealed with shellac, is shown with one side of the gelatin mold complete and the other partially begun. The second half is covered with a thickness of clay that will be the thickness of the gelatin. Gelatin will be poured through the knobs, seen in the photo, after the plaster support mold is made over the clay and the clay removed. A clay wall, which is dividing the upper from the lower section, will be removed when the lower section is completed.

The clay wall with temporary clay reinforcement. Note the clay ridge around the piece that will reinforce the outer edges of the flexible gelatin mold and secure alignment of the two halves.

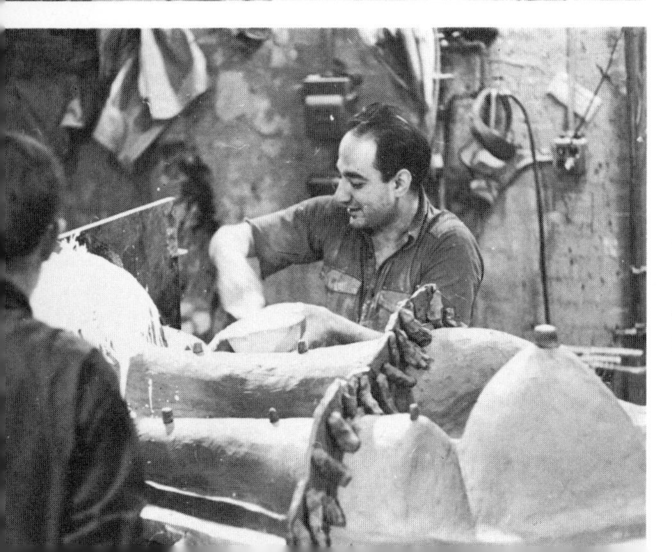

The plaster is applied to the lower section first, allowing the vents to protrude. After the lower section is completed, the clay wall dividing the top from the bottom is removed. The plaster edge is coated with wax, petroleum jelly, or green soap. The upper section is made against the lower half. Reinforcement steel bars have been used in all sections; burlap or other fibers may be used for reinforcement also.

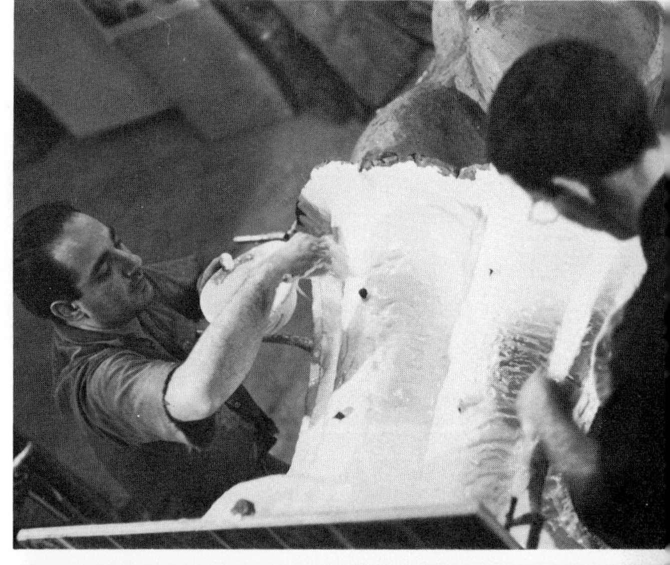

Another view of the plaster being applied to the lower section. The vents show as the dark marks.

Both the plaster and clay lower sections removed except for two knots of clay that protrude halfway into the upper section. When these knots are cleaned out they provide the holes into which the gelatin will be poured. The upper section will also be removed and the clay cleaned out. The entire mold will be reassembled and gelatin will then be poured into the space formerly occupied by clay. The gelatin is melted in a double boiler and allowed to cool to approximately 120°F and then it is poured.

A section of the plaster mother mold supporting the flexible gelatin mold. The first coat of wax is brushed in to assure good detail and to control the temperature of the wax. (If wax is too hot, it could destroy the gelatin.) The pouring cups, vents, and outer ridge help hold the gelatin in register.

One mother mold section; note pouring cups, vents, alignment ridge, and nipples.

The sections are clamped together with one coat of wax brushed on. Alignment is checked.

Molten wax is poured in at a low temperature so as not to melt the previous coat. It is allowed to set for a moment and poured out. This operation is repeated until the desired thickness of wax is achieved.

The complete gelatin mold with reinforced plaster mother mold wired together.

The wax castings are removed from the molds *after* they have been invested. A large section of investment extends from the core to align with the investment on the outside and seals the two sections together.

The protruding wax tabs will act as joints between the upper and lower sections when complete.

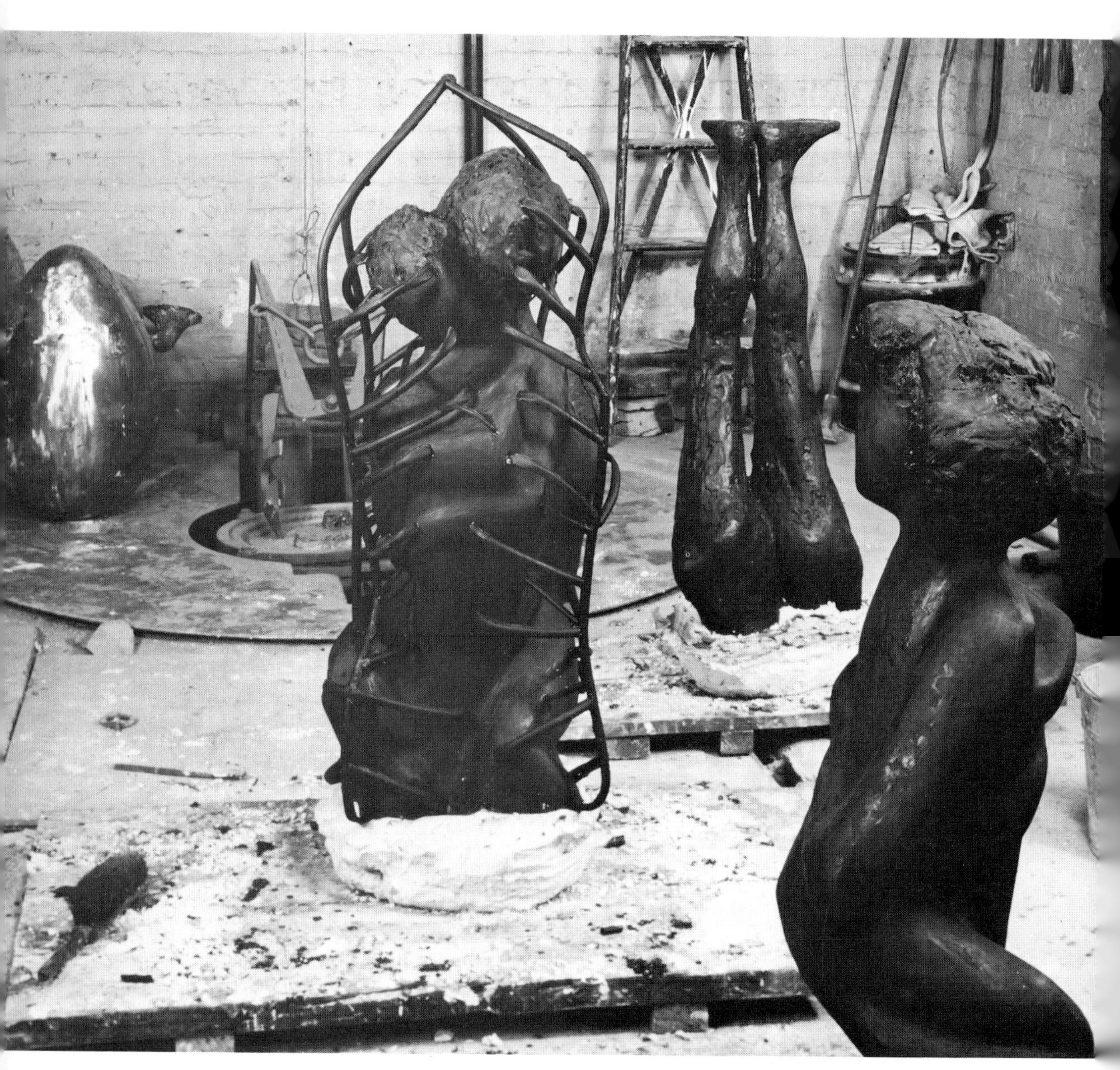

This different piece, MOTHER AND CHILD by John Kearney, illustrates another example of how the sprue system is attached before the wax is completely invested in the section at center of photo.

Both sections of AURORA shown with the lower section now completely invested. The investment was laid up by hand.

Detail showing pouring cup and vents.

The pieces are now partially submerged in an opening in the floor, and a gas-fired brick burnout is built around the molds. They are then burned out; the wax is melted and burned out and the plaster is calcined.

After the molds are burned out, they are packed within a framework filled with sand, which is rammed down around the molds. Should the mold explode or a leak develop, the sand will contain it. The pouring cup should be covered at all times to prevent foreign matter and cold air from entering the mold as they can cause poor castings and cracks in the mold.

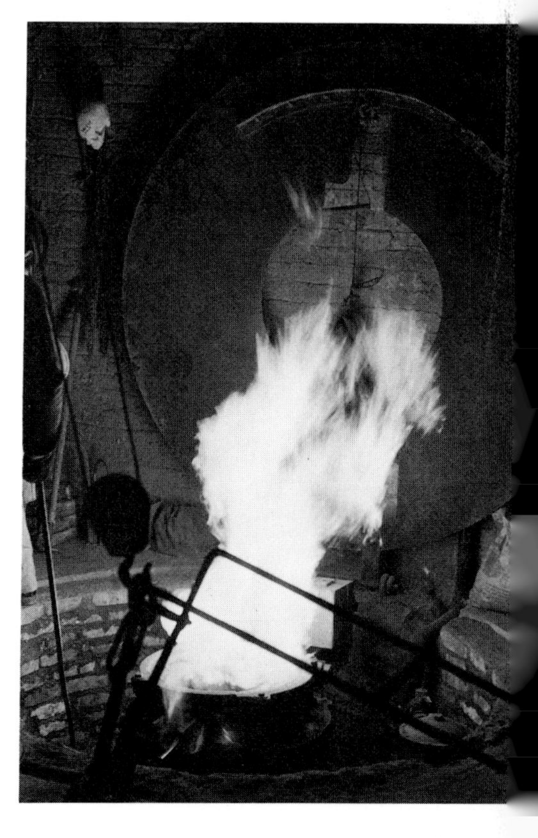

The gas-fired furnace is charged. The furnace can be melting the metal while the molds are being packed so that immediate pouring is possible.

Flux is dropped into the metal just before the crucible is removed from the furnace.

The crucible is lifted from the furnace with an electric hoist . . .

and bent handle tongs, and then placed into a two-man pouring shank.

The metal is skimmed of dross and poured.

Top and bottom sections devested. The sections will be cleaned and assembled. The tabs on the lower section will be drilled and tapped and screwed into through the top section. The entire seam will then be welded, ground down, and chased to blend with the surrounding metal, and the metal patinated with liver of sulphur.

A gelatin mold is being removed from the original and from the plaster mother mold.
Courtesy, The Morris Singer Foundry, England

TUNIC SERIES IV. Thomas Walsh. 1970. Silicon bronze using the ceramic shell mold, 26 inches high, 14 inches wide, 12 inches deep.

Courtesy, artist

8

The Ceramic Shell Mold

A CERAMIC shell, as its name implies, is created with raw materials of silica and binder that, when fired, become a homogeneous ceramic shell. Technically, ceramic shell casts are more sophisticated and recent than the sand or lost-wax methods of making molds. The method provides several advantages for small and medium-sized molds.

Ceramic shell casting offers incredible dimensional accuracy and surface detail—the main reason it has been developed and refined by industry, especially by Avnet Shaw in England. There also is less finishing required than with sand and lost-wax methods because fewer or no vents are used.

Ceramic molds, unlike investment and sand molds, are light in weight and highly resistant to thermo-shock, which allows longer burnout times. This also enables a variety of prototype materials to be used, such as paper or balsa wood, without fear of residue remaining unburned in the mold; such residue can be either poured, shaken, blown, or sucked out with air. In addition the mold can be removed from burnout to pouring immedidately, as the mold can withstand the thermo-shock.

Initially, the cost of building a ceramic shell setup would be slightly greater than other methods because of the materials for the ceramic shell and requirements of high temperature burnout. But the wax saved from runoff and the time saved from meltout through firing the shell, to pouring of metal and breakout of the casting would offset the cost. With experience the procedure can readily be accomplished in a good morning's work.

Despite the many advantages of ceramic shell casting, you should remember that the mold is more fragile or brittle than others; it is a ceramic shell that becomes hard upon firing. Once fired, any metal or alloy, ferrous or nonferrous, may be poured into the mold. Some smaller molds may be poured resting against another object while larger molds require outside support should pressure from the metal burst the mold.

The following ceramic shell casting demonstration has been developed by Thomas Walsh. Photos by John Tuska.

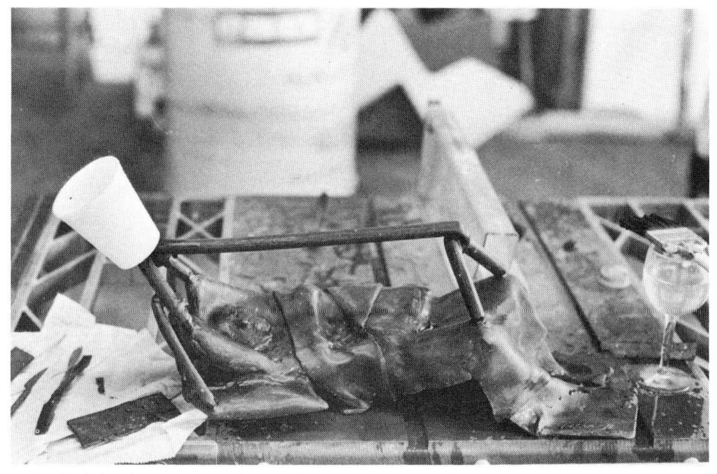

A finished wax model is being prepared for casting. The sprue system is attached and a foam cup is used for a pouring cup. Alcohol is used to clean oil from the wax surface and to ensure complete adhesion of the initial coat of slurry.

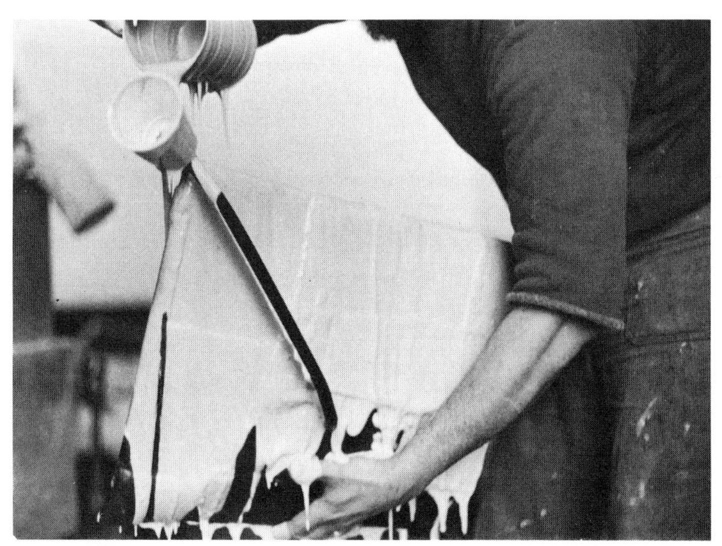

An initial surface coat of slurry is poured over the pattern. This coat of slurry must be a proper consistency since it picks up the surface detail. Consistency depends on the products you use; if you use a commercial ceramic such as Nalco's refractories, 1 part binder mixed with 2½ parts silica flour is very close to optimum. Allow excess slurry to drain off the pattern until there is a consistent thickness. You can also paint on the slurry or dip the pattern in a large batch of slurry.

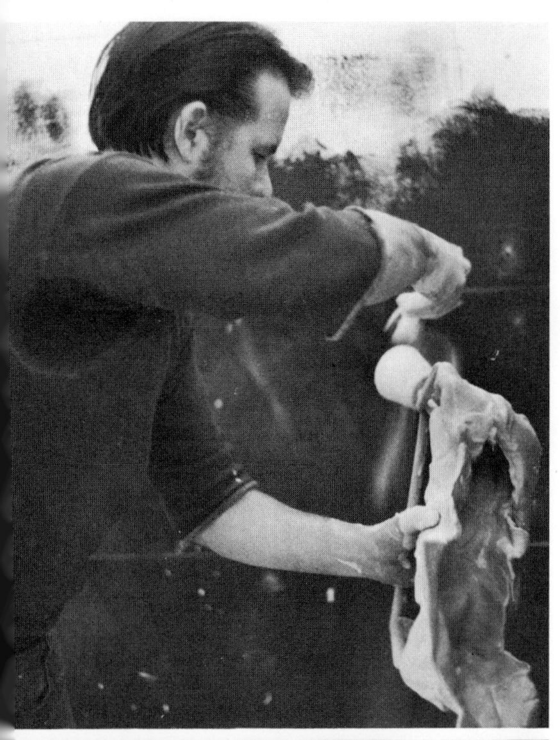

A stucco grade of fused silica is then sprinkled over the initial wet coat of slurry. This silica is very fine, which aids in picking up any fine detail on the pattern's surface. It also completely removes the problem of air bubbles. It is essential that the stucco silica be very dry when applied to the coating of slurry.

After the invested pattern has completely dried, the process is repeated until the desired shell thickness is obtained. Thickness is contingent on the size of the pattern and the metal to be used. These patterns were poured in silicon bronze at 2,050°F and the shells were approximately $3/16$ inch thick.

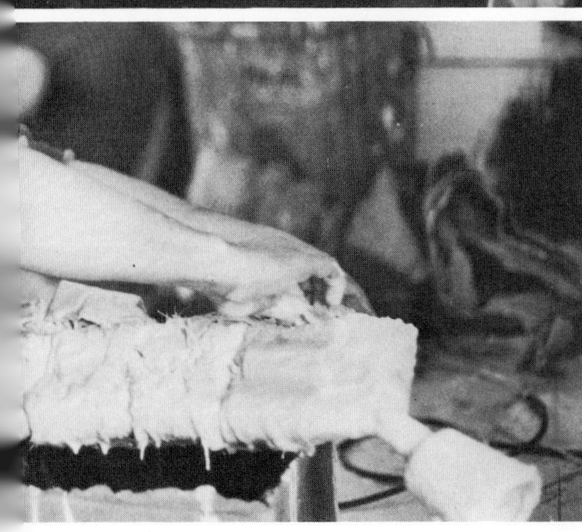

To reinforce ceramic shell molds, fiberglass cloth strips, dipped in slurry, are wrapped around the mold over areas that will need extra strength. With reliefs, reinforcement is especially important because of hydrostatic pressure that develops inside the mold against the extended mold walls, and the fiberglass impregnated with slurry is applied after the slurry and stucco layers are completed.

AIR DRYING SHELL

A group of pieces is set near a fan to accelerate drying time. Observe how different sprue arrangements have been used on the various pieces. During coating, the prime concern is to have enough slurry adhere for strength and to hold subsequent layers of fused silica. At times during the buildup of the mold the previous dried coating of slurry and fused silica may have to be "primed" with a coating of liquid colloidal silica before another coating of slurry can be applied, which assures an even coating of slurry.

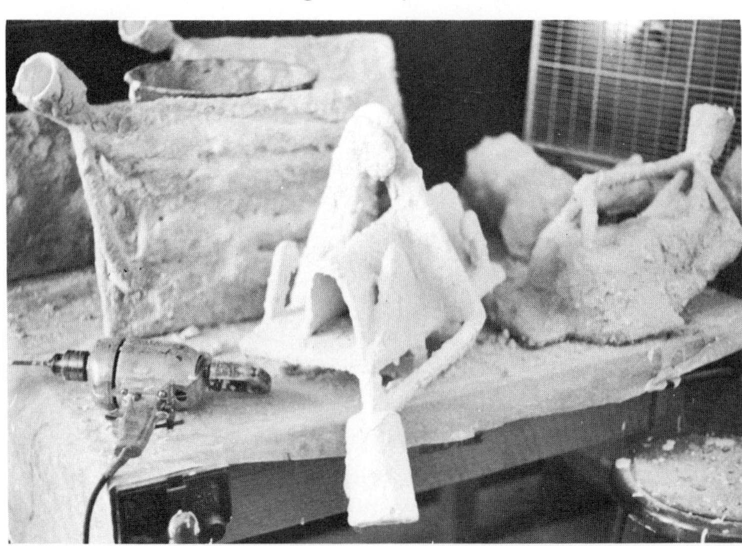

MELTING AND BURNING OFF OF ORIGINAL

The finished mold ready for burning off the foam cup and melting out the wax.

A portable burnout rack made from angle iron and insulation bricks is used in conjunction with two immersion burners to burn out the Styrofoam cup and the wax pattern.

The burners are ignited by hand.

The mold is hand-held to burn out the foam cup to start the melt-out.

A relief is supported on firebricks and the burner is directed at the pouring cup. The grate in the platform floor allows the melted wax running from the mold to flow through and be reused. (The wax may be used to make oil clay; it should not be used for lost-wax patterns since the carbon content from partial burning can be detrimental. A low-temperature melt-out will not generate a high carbon content.) With the foam cup burned away the wax begins to melt and the burner is moved gradually up the mold to melt out all the wax. This procedure helps to prevent excessive wax expansion and cracks in the mold.

HIGH-FIRING THE CERAMIC SHELL MOLD

The patterns are then placed in a conventional kiln and taken up to 1,700° F. This melts out any remaining wax and burns off all the carbon. Taking the shells up to this temperature allows the sculptor to pour the metal at a low temperature or "cold" and minimizes the gas and shrinkage problems. While the molds are in the kiln, the bronze should be melted so it can be poured when the molds are removed.

Close-up of ceramic mold burned out and fired.

CASTING INTO CERAMIC SHELL MOLD

Shell molds should not be rammed in wet sand. This would reduce their permeability and also cause a steam problem that would result in pitted castings. The investment patterns are set on a dry sand bed and the metal poured. The patterns are broken from the molds about a half-hour after they have been poured. Note the large screw on the pouring shank ring that prevents the crucible from falling out during two-man pouring.

TUNIC SERIES IV. Thomas Walsh. 1970. Silicon bronze made by the ceramic shell method, 26 inches wide, 14 inches high, 12 inches deep.

Courtesy, artist

STAINLESS STEEL ROLLABLE. Dennis Jones. This sculpture was first made in wood. A silicone rubber mold (RTV) was taken from the wood and then a wax cast was made from the RTV. The wax was then sprued and cleaned with denatured alcohol to remove oil. It was dipped into ceramic slurry and washed in silicone sand five times to build up the mold shell to the desired thickness. Burning out the wax followed; care was taken so as not to break the fragile green ceramic shell mold.

The mold was heated to 1,600°F and the metal, Hastelloy X (super alloy with high cobalt nickel content), was heated to 2,600°F in an electric furnace and the metal poured. As the metal congealed, the ceramic shell broke away. The sprues were cut off and the piece worked to a mirror finish.

Courtesy, artist

SPACE ROVER. Jacques Schnier. 1965. Bronze, 4 feet 10 inches high. See demonstration at right.
Courtesy, artist

Jacques Schnier developed an unusual process to cast SPACE ROVER. Foam plastic was used for constructing the sculpture which was molded in a ceramic shell for bronze casting.

The original sculpture is made from half sections of cylindrical ¼-inch-thick Styrofoam. Foam is light and fairly rigid but an armature is required for thin and extended sections. The piece, in foam, appears as it will in metal (opposite).

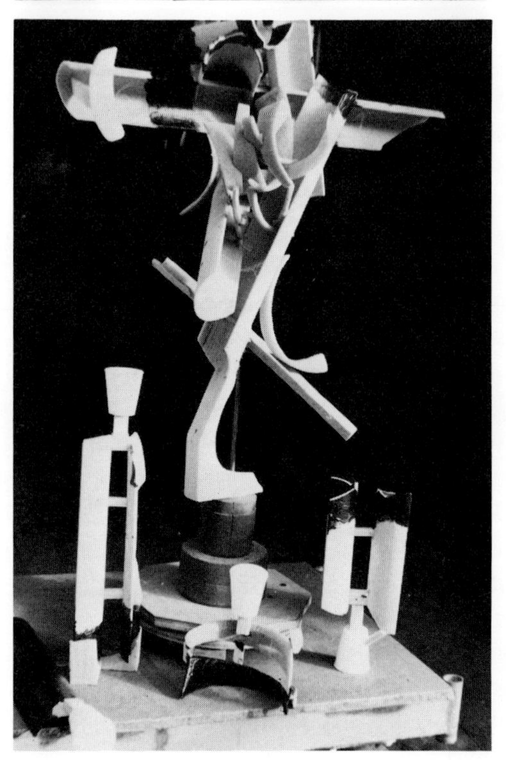

The 4-foot-10-inch sculpture was too large to be cast in one piece, so it was cut into sections that were coated with wax to achieve a different texture from the original foam. They were sprued, gated, and molded in a ceramic shell. Nalcoag Colloidal Silica (grade 1030) was poured into a container and Narcast fused silica (Nalcast P-1W) was blended in slowly with an electric drill until it was a thick creamy consistency. The piece was dipped in slurry, drained, and sprinkled with coarse silica grains. The process was repeated until the desired shell thickness was achieved.

When the ceramic shell was completed, the foam was dissolved from the mold with lacquer thinner. The mold was de-waxed (burned out) and fired as described in the earlier demonstration. After the pieces were cast, the sections were matched and assembled with heliarc.

Series courtesy, artist

9

Sand Casting

THE reasons a sculptor chooses one method of casting over another can be many and varied. They may depend on the equipment available or personal taste. Usually, however, they depend on the final end the sculptor desires. Sand casting offers certain advantages over lost wax when a high quality of the metal itself matters. With the use of fine-grain sands and newly developed binders, a more accurate surface detail than those that were available only a few years ago can be achieved.

Jack Zajac, for example, often prefers sand casting to achieve a highly polished, refined metal surface. He says, "In sand, the problem of porosity, a curse to mirror finish of bronzes, is largely eliminated because gases are permitted to escape through the sand particles during pouring of the metal. When these same gases are trapped within the denser walls of a lost-wax investment-type mold, they cause blemishes and spongy areas. The surface of an average sand-cast piece is not as accurate as that from investment castings nor is it advisable to have many undercuts; but for simpler forms, panels and pieces where the metal will be reworked, ground, and polished, sand casting is superior to lost wax."

Swan and Its Wake (see color page opposite 88) and *Falling Water* (opposite) were made using the following general procedures. The sculpture was cast in green sand using a core of sand. (Green sand is silica sand, 5 to 10 percent moisture, 10 to 20 percent bentonite clay.) First, a two-piece mold of sand was taken from the original plaster model. Then a sand image of the piece itself was reproduced by packing sand into the mold reinforced with a steel armature. This image was shaved or filed slightly (about $\frac{3}{16}$ inch) and was then suspended inside the sand mold, now acting as a core in its slightly reduced size. One pouring was required for the one-piece form, the metal being almost the thickness of the amount of sand removed from the core.

Shaping and finishing a polished piece is the most critical phase, even with good castings. To achieve the refined passages and good resolutions of form, especially on long spans, is exacting and intense work even with power and hand tools. There are few moments of mindless labor: it seems that every stage is crucial, from the cutting off of the sprues to the application of jewelers' rouge. The piece is ground down with the whole range of disks on an electric sander from rough to fine, and then a foam-backed pad is used to feather out any grinding marks or steps. Hand sanding follows with wet/dry paper grits 250 to 600. Buffing with rubbing compound and then jewelers' rouge completes the cycle.

One of the most difficult refinements to achieve is clean, unwavy surfaces on a large simple or flat span of metal. Only the hand, at times, will reveal ripples that can then be rough polished and a slight patina induced by applying liver of sulphur in water. After another sanding, the irregularities will appear as contrasting clean areas to patinated areas.

Sand casting is a relatively cheap and versatile method for producing art castings and has the advantage of not requiring expensive equipment. As the name sand casting implies, the molding material has sand as the main ingredient. Sand is defined as small particles of matter usually 0.1 to 1 millimeter in diameter. The sand may be silica, zircon, olivine, chromite, straulite, chamotte, etc. Finer sands give finer finishes and vice versa. Cleaner sands are more controllable. Round grains are generally preferred.

The sand is mixed with a binder and then rammed around the pattern. Several types of binders may be used. For each binder there exists a range of special additives used to produce special effects. Some of the more popular binders are:

1. Clay bonded
2. Resin bonded
   - A. Air set
   - B. No-bake
   - C. Thermal set
3. Sodium silicate
   - A. CO_2 set
   - B. No-bake
4. Cement

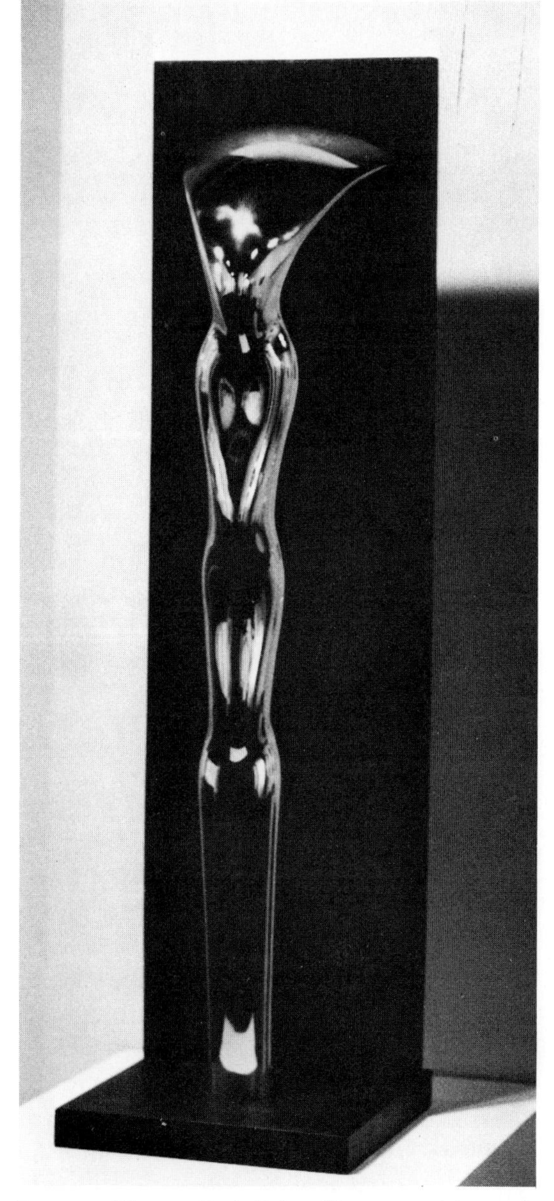

FALLING WATER. Jack Zajac. Bronze cast in sand, 8 feet high.

Courtesy, artist

Clay Bonded: Three to 18 percent clay is mixed into the sand with 2 to 10 percent water. Some kind of mixing device is usually required. Inexpensive hand-operated equipment may be obtained. The resultant sand, usually called a green sand, is stiff and must be "rammed" against the pattern. The pattern is then removed and the mold can be poured quickly. Special additives can be used, for example:

Corn flour for plasticity and thermal stability.
Sugar for some early strength while pouring.
Sea coal for development of a reducing atmosphere.
Iron oxide for development of hot strength.
Vinsol resins for cleaner castings.

Resin Bonded: Resins are generally man-made products and, therefore, can be tailor-made to do certain jobs. In general,

General equipment for processing synthetic green sand molds using a 55 grain size clean sand, 15 to 20 percent combined fireclay, and bentonite with 5 to 10 percent moisture.
On the wall, left to right: safety goggles, pickup tongs, molders, bellows, shovel, skimmer, and sprue cutters. On the floor: aluminum flasks, parting dust and bulb with brush, ingot mold, aluminum and brass flux, asbestos gloves, foundry riddles, molders' leggings and apron, buffing compound, wood ramming tool, set of trowels, a gate cutter and spoon, slick and lifter tools, and a full-face mask.

the resins are liquids that are mixed with the sand (about 2 percent or less) and allowed to harden, after which the pattern is removed. Further hardening is usually required before pouring. Resins are easier to use than are clay-bonded mixes and may be added or carved into for more detailed work or repairs. Resin-bonded molds may be hardened by air, chemical additions (the no-bakes), or heat (mold placed in an oven at 200° to 550° F.). Additives include iron oxide, flake resins, acids, sea coal, and sugar.

Sodium Silicate: Sodium silicate is a kind of inorganic resin. It is a liquid that is mixed into the sand (about 4 percent or less) and hardened, after which the pattern is removed and the casting poured. The sodium silicate systems are safe and easy to work with. The hardening may be accomplished by passing carbon dioxide gas through the mold. Another method consists of mixing certain chemicals into the silicate-sand mixture that cause the mix to harden after a half hour or longer. Special additives consist of sugar, iron oxide, and sea coal.

Cement: Regular portland cement can be used with sand to make molds. The retained water can be a problem, but good castings can be produced cheaply.

Today the trend in industry is to use premixed molding sand or sand additives.

Green sand casting is illustrated in its most primitive form, using the simplest available materials, in the following series of photographs taken at the Hinckley-Haystack School of Arts and Crafts. Plain washed river sand is hand riddled or sifted through a fine wire mesh to obtain an approximately even grain size. The sand is then mixed with 15 to 20 percent fireclay and bentonite. The fireclay increases hot strength, which holds the sand particles together as the hot metal contacts it. The bentonite increases green strength and permeability. The sand and binders are dry mixed in a portable cement mixer and 5 to 10 percent water is added as a binder; this is called tempering the sand mix.

Self-setting resin-bonded sands are used very successfully in cases where molds are made in pieces and the fact that the sand is air setting eliminates the burnout and mold baking processes. Self-setting resin-bonded sands were used to cast *Ominous Ikon* by Dennis Kowal (page 160) and *Etude: Apocalypse* by Caroline Lee (page 174).

The carbon dioxide (CO_2) gas process uses a sodium silicate bonded sand which is a mixture of approximately 3 to 4 percent sodium silicate mixed with clay-free silica sand. The mix is packed and rammed around the pattern. CO_2 gas is then forced throughout the mold under low pressure, a process called gasing the mold. From a tank of compressed CO_2 a perforated metal tube can be thrust into the sand for gasing or by attaching a hose to a flexible cup that can be held against the mold's walls which forces the gas into the sand, a perfect solution on small molds. For large molds, metal or plastic tubes may be placed into the mold during the ramming and the gas can then reach all the deep areas of the mold. CO_2 gas is applied for approximately 20 seconds, but the size and thickness of the mold will be the determining factor. For finer detail on the casting, a refractory mold wash such as zircon flour suspended in alcohol may be sprayed on the surface of the pattern. Two to three coats are applied and, upon drying, the piece is packed.

CASTING BRONZE WITH SELF-CURING RESIN SAND PIECE MOLDS

The following demonstration illustrates how one of eighteen sand mold sections is taken from the original plaster piece, *Ominous Ikon* by Dennis Kowal, and then cast in bronze, which is the culmination of two and a half years' work.

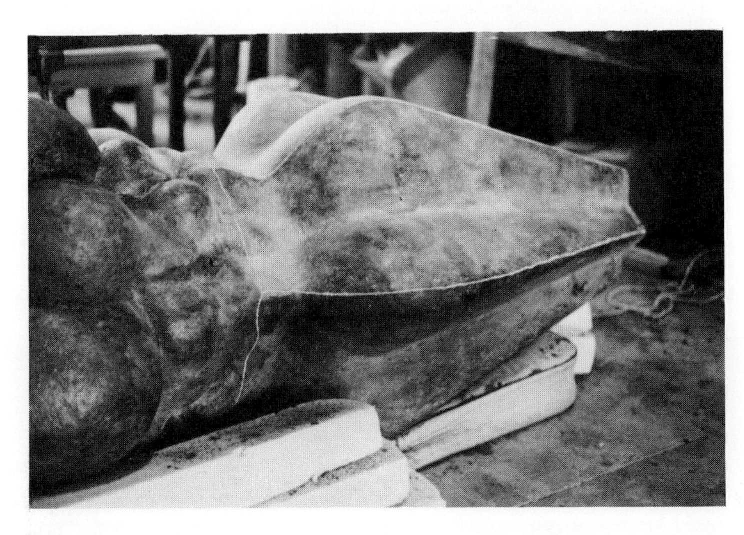

The plaster original of OMINOUS IKON is resting on its side with incisions indicating where the various mold sections will be divided. The walls that will contain each section of sand overlap the incisions that, in turn, will mark the sand to indicate each section.

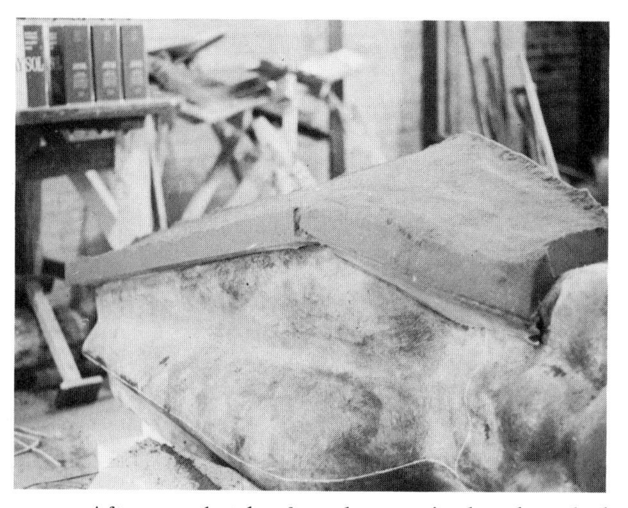

After one batch of sand was mixed and packed into one section and cured, the walls were removed and this shape remained. A two-part resin was used and mixed, or mulled, with the sand in a converted cement mixer following manufacturer's directions. However, the blades from inside the cement mixer were removed and ingots or round stones were thrown in with the sand and resin to provide the proper mixing or mulling of the resin and sand.

The first section was removed from the piece and the edges cleaned.

The section was marked for later identification and an area was added to accommodate a section of metal that would later be drilled and tapped for anchoring the finished piece. This was done on all lower sections.

One-quarter inch of clay was laid over the entire inside surface of the outside section of the mold. The clay is laid to the incision.

A wood frame was built around the section and the entire surface was dusted with fine talc to act as a release agent. Talc is distributed evenly by using a mesh fabric bag.

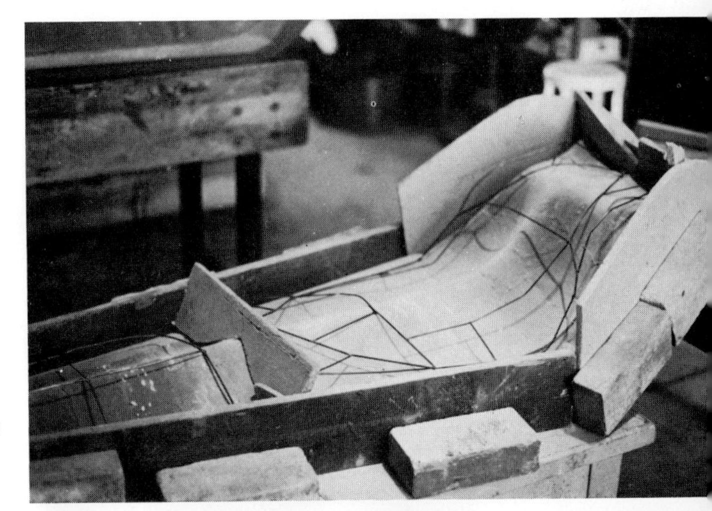

A metal reinforcement was fitted to the shape of the mold.

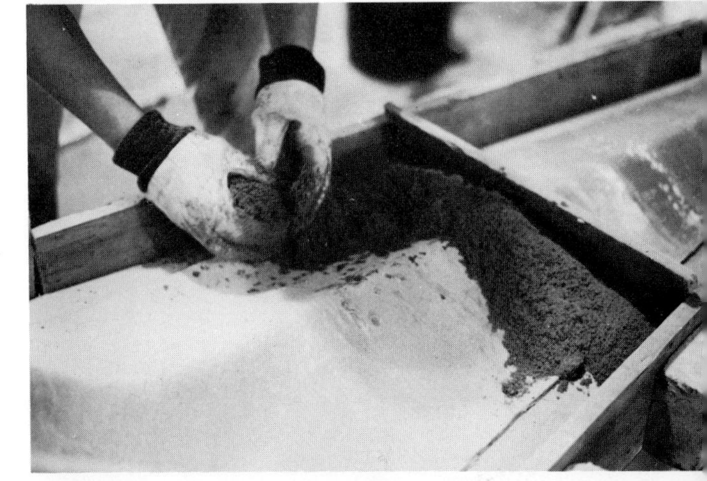

The prepared sand is being packed within the frame on the clay.

The wire reinforcement was packed within the sand. The setting time of the resin varies according to temperature and the resin-catalyst ratio used. Average *working* time is ½ hour and total curing time is 2 hours.

The first of two inside sections is now complete, and pouring cups and vents are laid out.

The first part of the inside section was removed after the second section was made against it; the clay was removed leaving a ¼-inch space between the inside and outside sections.

Here, you can see the gates and vents cut into the inside sections. A coating of refractory core adhesive is applied to the mold section edges. Observe the alignment cuts on the outside edge of the mold.

►

The edges of the outside section are also coated with adhesive. The surface of the mold was coated with an alcohol and graphite mixture that is flamed or burned off; this increases the mold's surface strength against metal erosion, which causes particles of sand to remain in the casting.

The two sections are glued together; care must be taken not to get adhesive on the mold's surface.

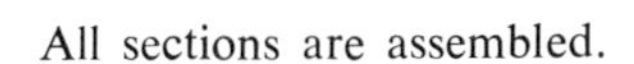

All sections are assembled.

Pouring cup and vents are glued in place; all are made large enough to allow for shrinkage as the metal inside the mold cools. Use a commercial adhesive or mix together clear lacquer, red iron oxide, and silica flour as the glue.

Painting refractory core adhesive on pouring cup.

The pouring cup is aligned before the adhesive hardens.

The mold is now complete. Notice the *small* vent holes.

The mold is banded with metal strapping. Wire or clamps may be used also.

Seams are sealed with a clay and sand mixture as an added precaution against leaks.

The metal is poured. Metal and gases vent from the lower openings. Two men pour while another stands ready with a shovel of sand to cover vents and breaks should they occur. After the metal has cooled, the section is removed from the sand mold and the first of the eighteen sections to be cast is completed. See OMINOUS IKON, next page.

OMINOUS IKON. Dennis Kowal. Bronze, 8 feet high. The finished piece.

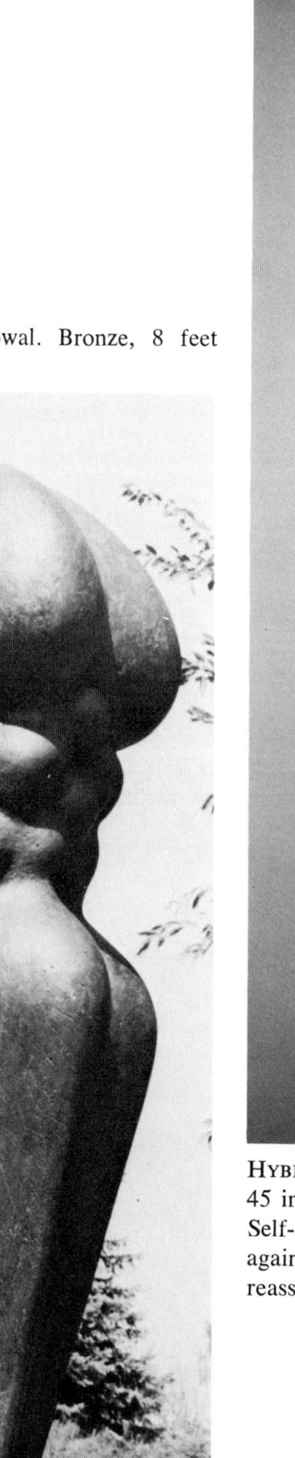

HYBRID FORM #1. Richard Hunt. Cast aluminum, 45 inches high, 16 inches wide, 12½ inches deep. Self-setting resin-bonded sand packed directly against clay original in sections; removed, then reassembled and cast.

Photo, O. E. Nelson

AMERICAN QUEEN. David Hostetler. Bronze with an ammonium sulphide patina, 58 inches high. Resin sand casting.

Courtesy, artist

UNTITLED. Julius Schmidt. 1969. Bronze, 51 inches high.

Courtesy, Marlborough-Gerson Gallery, Inc. New York

The accompanying series illustrates the development of the sculpture *Memories of Bali* (page 164) by Jacques Schnier, using a foam vaporization technique for casting a hollow bronze in sand.

The core of the sculpture composition is built of odd blocks of core sand held together with a core adhesive made of high temperature materials such as mullite, zircon, fused alumina, or magnesia.

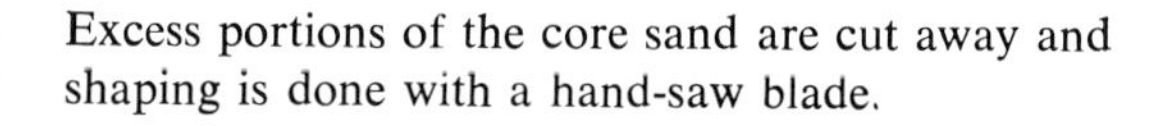

Excess portions of the core sand are cut away and shaping is done with a hand-saw blade.

The finished core. Irregularities are filled with refractory core paste.

The finished core form is covered with ¼-inch sheets of plastic foam (Styrofoam) cut to fit the configuration of the surfaces. The top section has been removed for ramming in sand; it is packed in sand separately.

Foam sheets, ¼ inch thick, are bent to cover noncompound curved surfaces.

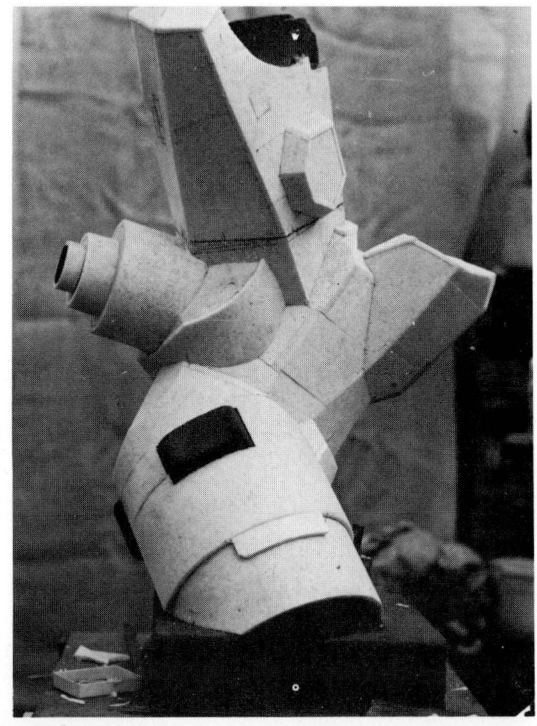

Bottom section has runners, gates, and small risers attached. Surface is brushed with a wax filler and then with a refractory mold wash to reduce roughness. The piece is carefully placed in a flask or wood container and sand is rammed around it. The metal is poured, burning out the foam as the form is filled. The sections that were cast separately were then assembled with heliarc welding.

Photo series courtesy, artist

◀

MEMORIES OF BALI. Jacques Schnier. 1964. Bronze, 4 feet 2 inches high, 22 inches wide.
Courtesy, artist

▶

UNTITLED. Julius Schmidt. 1961. Iron, 21½ inches high.
Courtesy, Marlborough-Gerson Gallery, Inc., New York

ENCOUNTER IV. Lynn Chadwick. Bronze.
Courtesy, The Morris Singer Foundry, Ltd., England

SINGLE FORM. Barbara Hepworth. Original plaster. Dag Hammarskjöld memorial for United Nations Building, New York.
Courtesy, The Morris Singer Foundry, Ltd., England

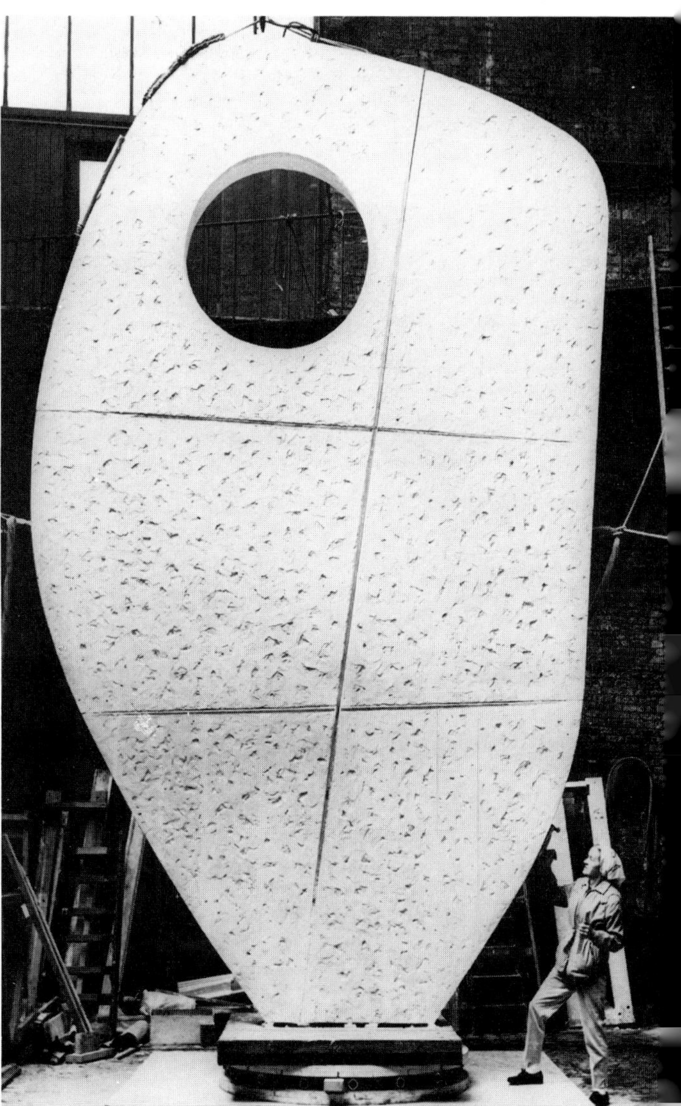

Sand molding shop.
Courtesy, The Morris Singer Foundry, Ltd., England

Sand casting—replacing core into mold.
Courtesy, The Morris Singer Foundry, Ltd., England

YOUTH #2. Howard Woody. Cast aluminum, 76 inches high, 72 inches wide. Resin-bonded sand mold, hollow. Twelve pieces, heliarced and bolted.

Photo, Elliott Borenstein

Land Monument. Nancy K. Sprague. Bronze and wood, 27 inches high, 21 inches wide, 13 inches deep.

Courtesy, Ruth White Gallery, New York

Primavera (Study). Lawrence Fane. 1968. Bronze, 14 inches high. Sand cast and finished by filing and hammering. Sections are polished to a mirror finish.

Courtesy, Zabriskie Gallery, New York

EPISODE G. Edgar Tafur. 1970. Bronze with chrome finish, 14 inches high, 18 inches wide. Sand cast from wood patterns.

Courtesy, artist

UNTITLED. Martin Chirino. 1970. Bronze.

Courtesy, Galeria Juana Mordo, Madrid

UNTITLED. Richard Kowal. 1968. Aluminum, 36 inches long, 30 inches high.
Courtesy, artist

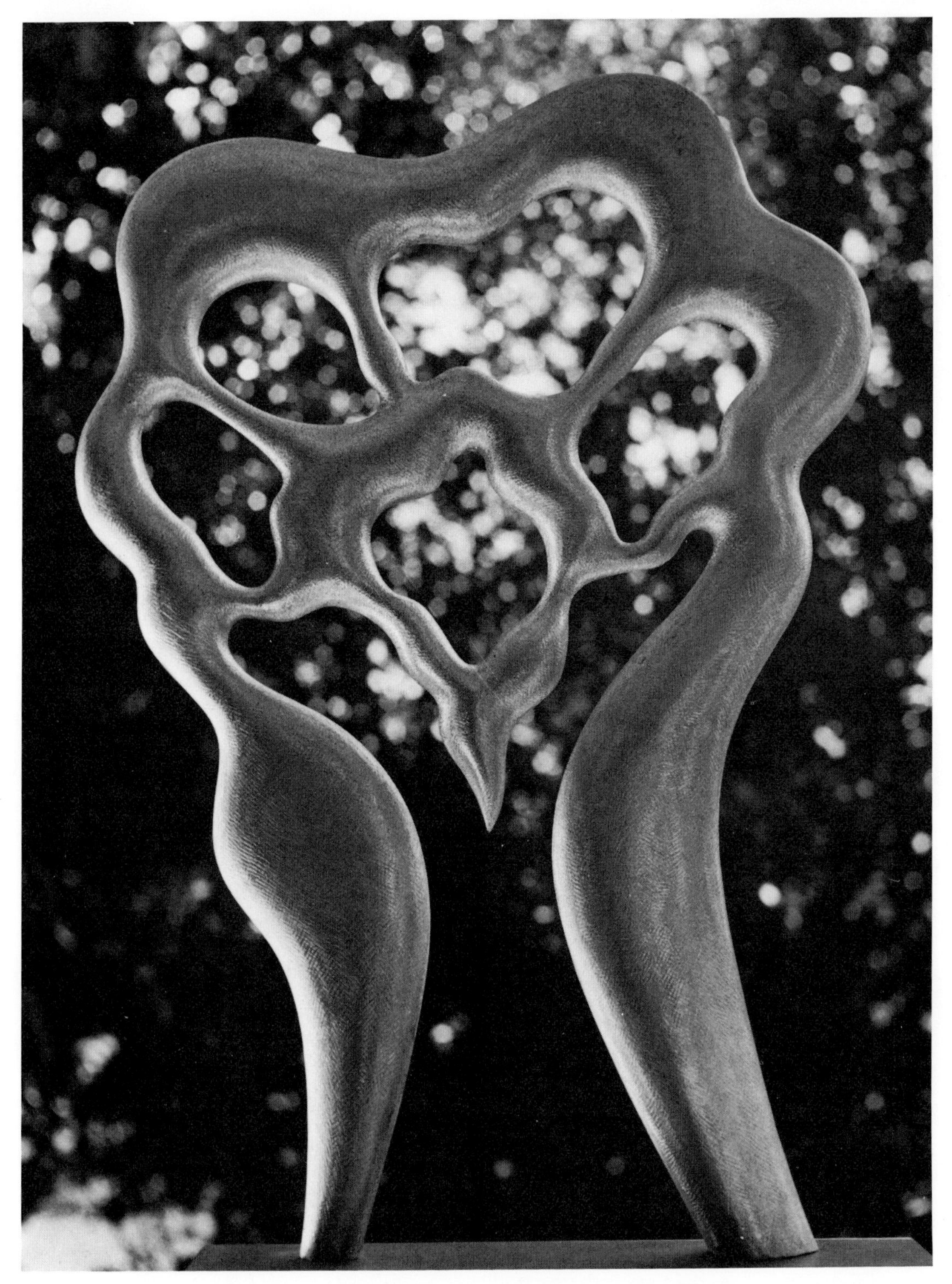

HOMAGE TO ARP. Dennis Kowal. Sand cast aluminum from wood, 27½ inches high.

◄

VOLUME II. Gottfried Honegger. 1968. Cast aluminum, edition of four, 14¾ inches high, 6 inches wide.

Courtesy, Gimpel & Hanover Galerie, Zurich

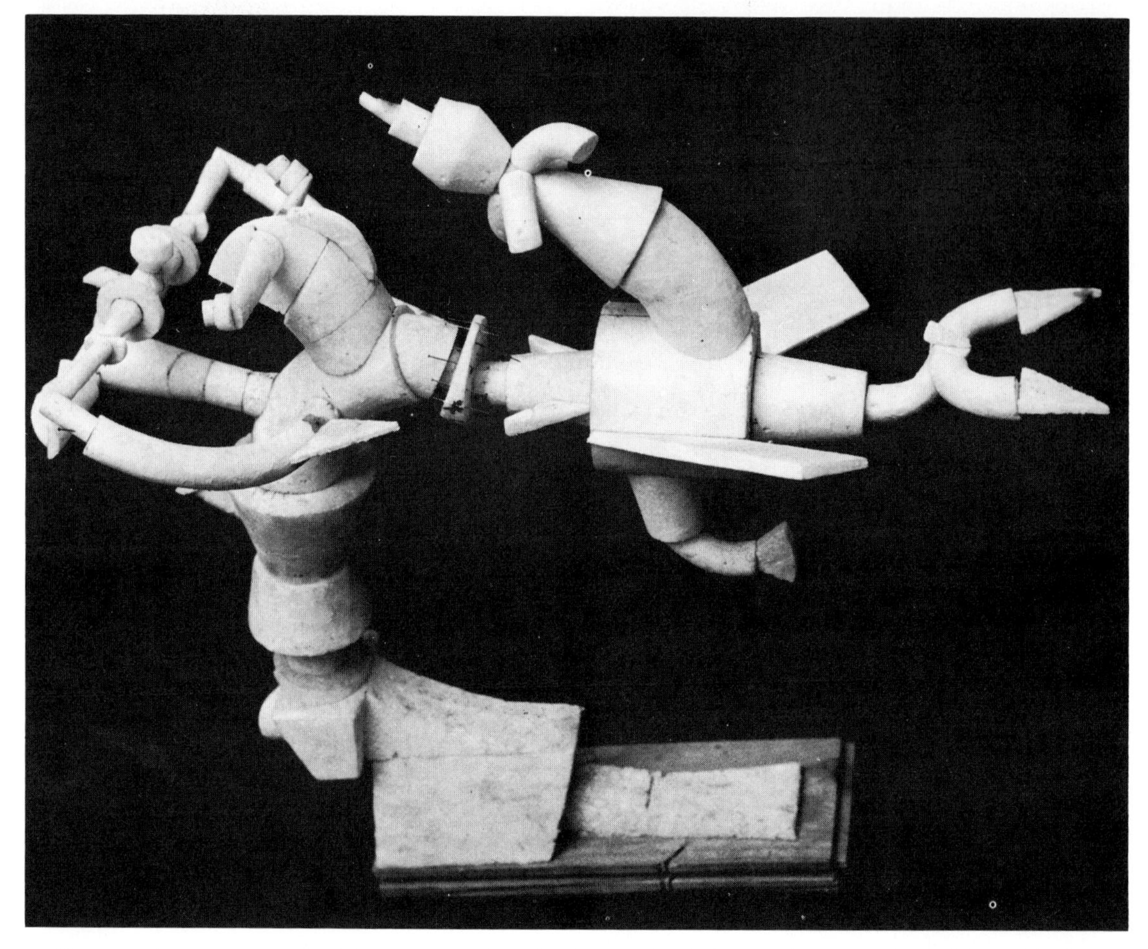

ETUDE: APOCALYPSE. Caroline Lee. 1969. Aluminum, 50 inches high, 50 inches wide, 35 inches deep.

Photo, Augustin Dumage

LOST-FOAM METHOD FOR CASTING IN OIL-BONDED SAND

Caroline Lee of Paris illustrates the procedures she follows for casting metal using the lost-foam method in sand. In this technique the foam pattern is lost during burnout. Photo series, Augustin Domage, Paris.

A finished piece of sculpture is ready for the gating system. A low-density foam is used and some parts are held together with wire and others are glued permanently. The wire sections can be taken apart and cast separately. The foam can be shaped with saws, files, and, sometimes, with a hot wire or heated tool. Sections may also be dissolved with various solvents such as lacquer thinner. The foam may also be coated with wax for a smoother casting and then with a refractory wash for a smoother surface.

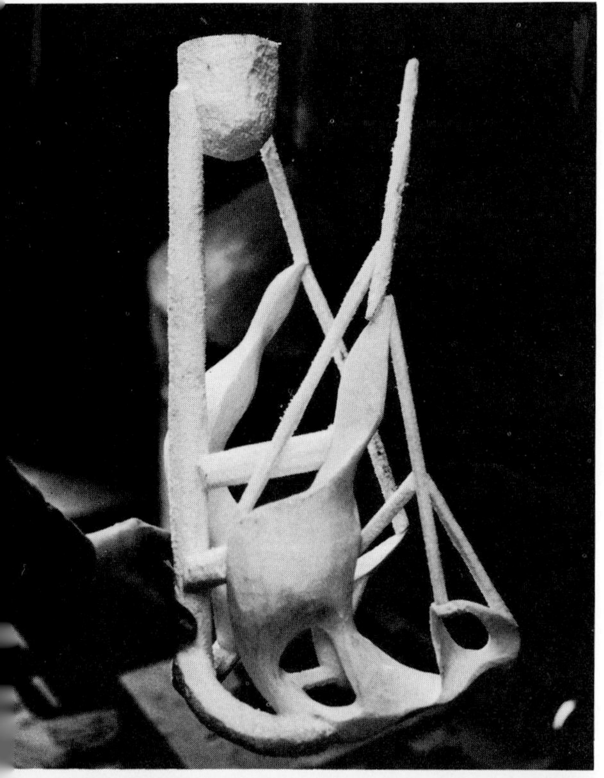

A foam piece with the gating system attached by glueing. Sometimes Miss Lee glues only the beginning of the vents to the piece, continuing it in the molding process by a length of hose that can later be withdrawn from the finished mold. This allows a completely free passage for the escaping gases emitted by the foam, binding oil, and metal. As the foam is vaporized it seems to form a vacuum into which the metal is drawn. The result is a good casting despite deep undercuts that are hard to pack with sand.

A fine-grain silica sand is used, mixed in a muller with a commercial polymerizing oil and catalyzer called Petro-Bond (a common waterless sand binder for ferrous and nonferrous metals); a blended oil is sometimes used to improve the temper of the sand along with the Petro-Bond. Wooden flasks of various sizes are used to pack the sand around the pattern, which allows the sand to be the proper thickness around irregular sections and also saves sand.

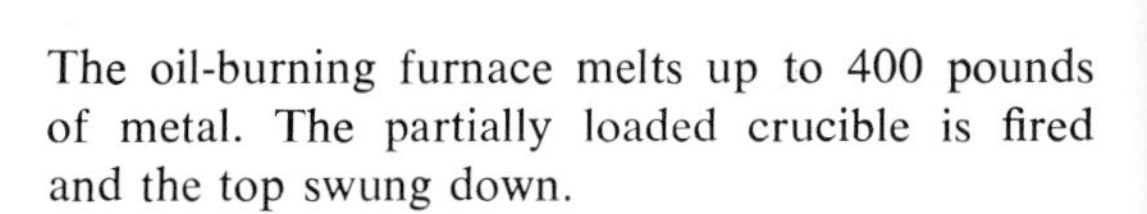

The oil-burning furnace melts up to 400 pounds of metal. The partially loaded crucible is fired and the top swung down.

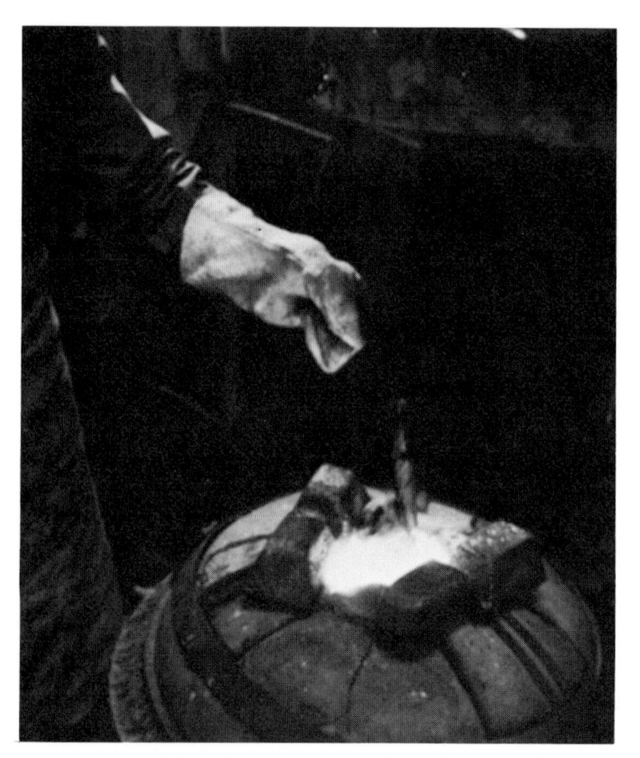

Additional ingots are preheated on the furnace top before they are lowered into the furnace as any moisture on the metal could cause an explosion inside the furnace.

Deoxidizing the bronze immediately before pouring.

The crucible is removed from the furnace, the metal skimmed and poured. Observe the flames and gases escaping from the vents. The crucible is a two-man shank with a hoist assist. The metal must be poured as quickly and as hot as possible or else it will not sustain itself for a good burnout of the foam and a poor casting will result. The mold should be broken open as soon as possible as the hot metal will bake the sand and make its removal more difficult. Note, however, that though aluminum may be solid in appearance it is very weak in the transition from a liquid to a solid; bronze less so.

►

THE SHOUTER. Caroline Lee. 1965. Bronze, 1 meter, 40 centimeters high. Cast from foam in parts that were bolted and welded together.

Courtesy, artist

HAZZOS HORN. Marie Taylor. Yellow brass, 18 inches high.

Collection, Mr. and Mrs. James Singer
Courtesy, Betty Parsons Gallery, New York

LITTLE JOE. Cabot Lyford. Bronze, 9 inches high. Cast by the lost-foam method.

Courtesy, artist

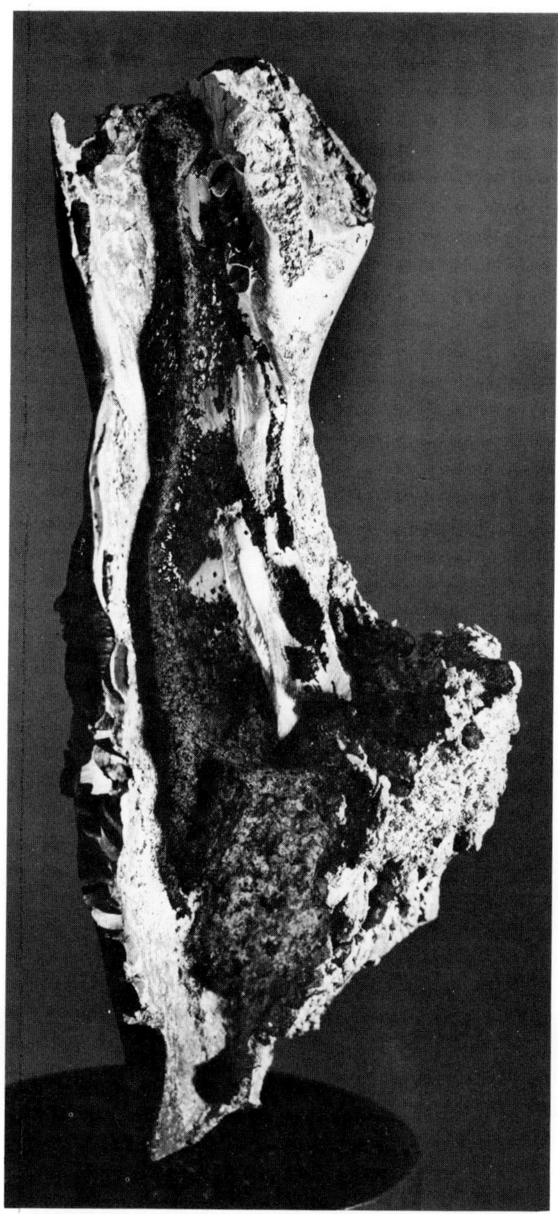

UNTITLED. Jan Zach. 1960–62. Cast aluminum, 14 inches high.

Photo, Ken McAllister

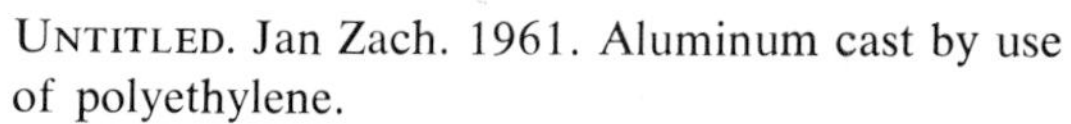

UNTITLED. Jan Zach. 1961. Aluminum cast by use of polyethylene.

Collection, Museum of Art, University of Oregon
Photo, Ken McAllister

THE MELT-OUT PROCESS FOR ALL COMMON METALS

The melt-out process and the use of a cupola furnace, both usually associated with industrial iron foundries, has not often been employed by sculptors. However, Julius Schmidt refined the melt-out process while he was at Cranbrook Academy, Michigan, for casting sculpture, and Stephen Daly shows how it is done in the following series of photos taken at the "Sculptor's Foundry," Humboldt State College, Arcata, California.

The melt-out process is an inexpensive method of making metal castings from patterns and offers several advantages over investment castings. Because there is no plaster to react with the metal (as in standard investment casting), there is never any fire scale on the castings. Silica sand, which is less expensive than plaster, is the main composition of the mold. No elaborate kiln is required as the molds are simply baked at a low temperature which conserves fuel. Either CO_2 or a self-setting catalyst resin sand is packed around a pre-coated wax pattern. The sand and the Ceramol 55 ceramic coating form the mold. No surface defects occur, such as the flashings or bubbles associated with investment casting. Any metal from magnesium through Monel can be poured in this type of mold.

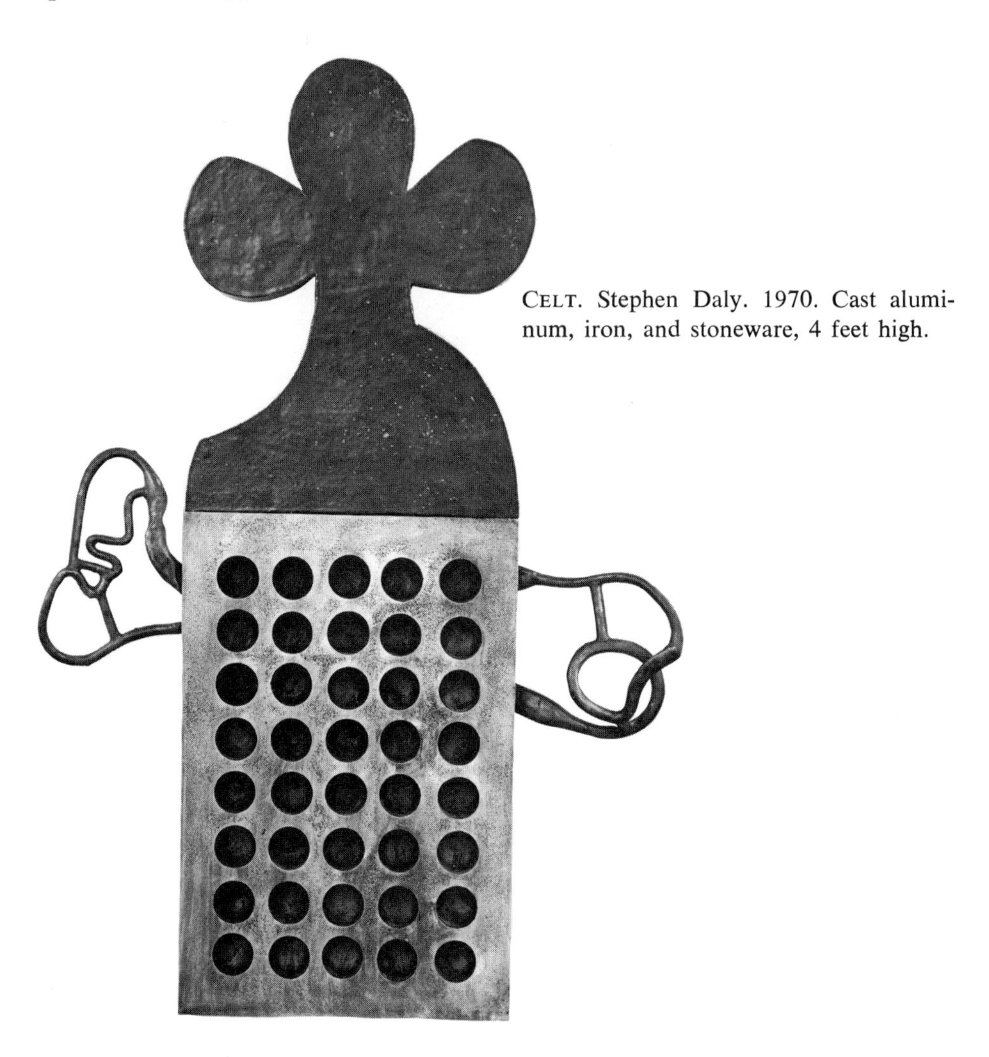

CELT. Stephen Daly. 1970. Cast aluminum, iron, and stoneware, 4 feet high.

Piece to be cast is spray coated with Ceramol 55 (a patented Foseco product). Pattern can be microcrystalline wax, fabric, or other pattern material. Wall thickness for bronze should be about $\frac{3}{16}$ inch; for iron, a little thicker. Sprues and vents should be slightly larger than those used in investment casting. The coated wax pattern is rammed into self-setting resin sand and then placed upside down in a kiln for approximately 12 hours at 350°F; thermocouples are used to monitor temperature. Some of the wax may be caught and saved but most of it is absorbed by the sand.

The mold, drained of wax and baked, is packed and rammed in flasks or frames. The pouring openings are protected with aluminum foil. The Ceramol maintains its accurate impression of the pattern and will keep the metal from penetrating the sand.

Student keeps a record of the progress of the melt in the cupolas.

IRON POUR CHART

USING 12-INCH (APPROXIMATELY) INSIDE DIAMETER CUPOLA

Fire Started	11:55	Each Charge Contained:	
Bed Burned In	2:55	Iron	75.5#
Metal Charged	3:12	Coke	12.6#
Air On Full		Limestone	1.7#
Tap and Slag Holes		Calcium Carbide	
Botted	3:13	(Producing Acetylene)	
Drops	3:17	2# On Bed	
		2# First & Second Charges	

See also Appendix, page 252

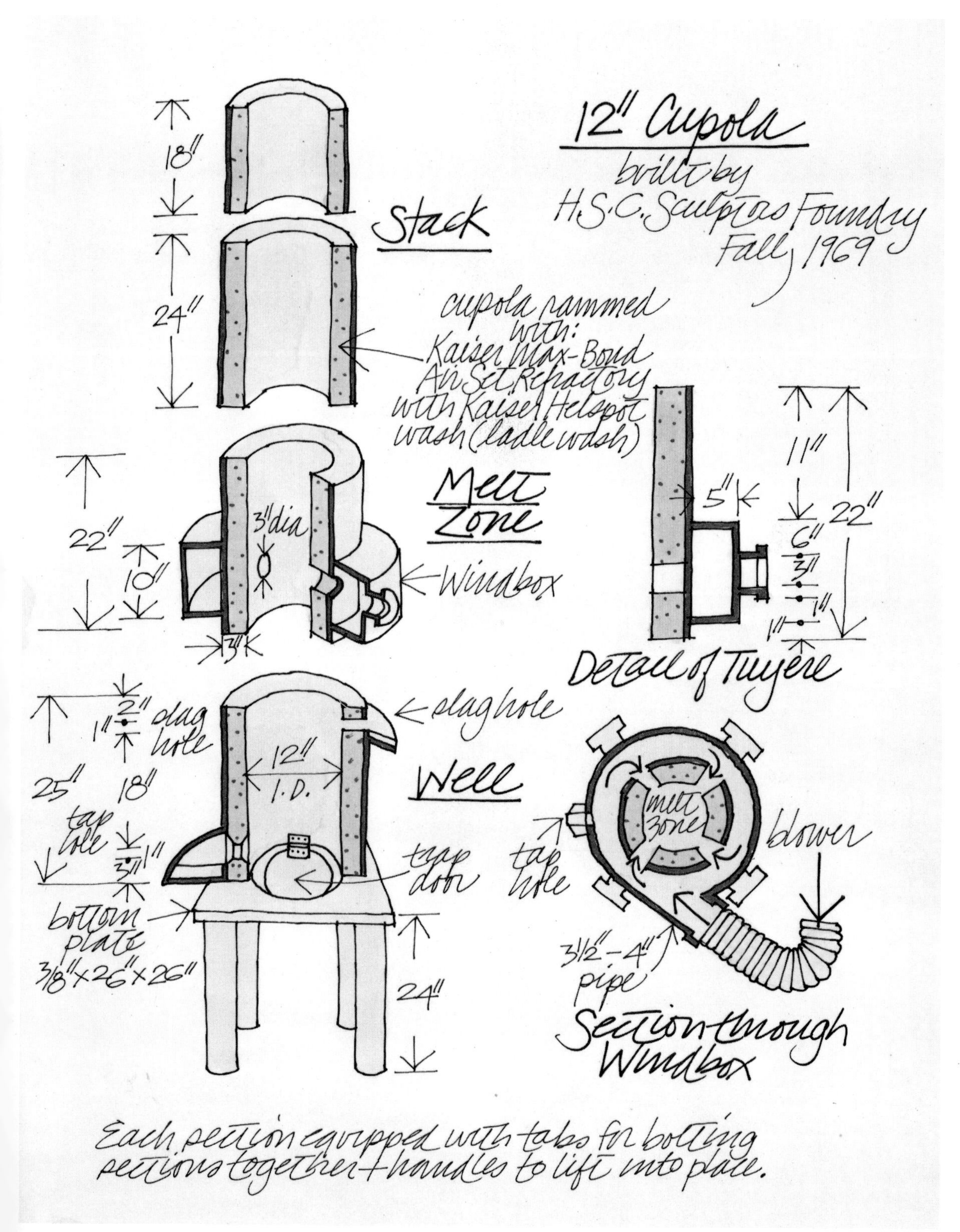

Structure of a 12-inch cupola furnace.

Drawing, Lebbeus Woods

The metal that is trapped at the bottom of the cupola must be "tapped out." Here, a round chisel is used to penetrate the bott that is holding it back. The bott is made of approximately 75 percent fine clay and 25 percent coal dust or sawdust and a small amount of water.

After the crucible is filled with metal, Daly places another bott in the taphole of the cupola to stop the flow of iron.

Five people are involved in this pour: two are skimming, two are pouring, and the fifth is operating the hoist that assists with the 500-pound ladle. The heat from the molten metal is dissipated to the mold which breaks down the sand binder and the piece is easily removed from the mold.

IRON SCULPTURE. Julius Schmidt. 1959. Cast iron, 19½ inches high, 37½ inches wide, 19 inches deep.

The Art Institute of Chicago

The work of Julius Schmidt epitomizes the experience of a man whose influence and knowledge has furthered the development of various sand casting methods. In one such development Schmidt begins a sculpture by carving in reverse, using cured or baked core sand blocks that are reassembled and identified. He then makes a system of cores to fit within the carved blocks, so that the castings will be hollow. The pouring cup and sprues are cut into the sand before the pieces are glued together. When he uses wax for the original pattern it is usually melted out in a core oven or kiln while the sand is set. Schmidt's pieces are cast in bronze or iron.

LEAD CASTING

Lead castings, unlike bronze, can be made without any special equipment. In fact, small pieces up to about 30 pounds can be cast using only a gas cooking stove for a heat source. Lead, pewter, and some of the newer low-temperature alloys are excellent casting materials that do not require major equipment. Because they are relatively soft, they can be reshaped after they are cast. In addition, lead is resistant to the corrosive atmospheric conditions that exist in our cities today. It has esthetic potentials that have not been adequately explored.

The mold required for casting molten lead can be made of molding plaster that has been calcined, that is, the water has been driven off. This is accomplished by baking the mold in a cooking-stove oven at its highest temperature for 12 hours. When a wax original is used, a pan of water should be set under the piece in the oven and heated at 150° to 200° until all the wax is melted out; then proceed to calcine the mold. Drain the wax from the pan often to prevent fire. It is possible to build a temporary brick structure on top of the stove resembling a kiln and following the same procedure.

To melt the lead, a steel plumbers' ladle is used, but a steel cooking pot will work as well, depending on the amount of lead to be poured: larger amounts require more sturdy equipment. A brick rest can be built around the largest burner; an acetylene torch can also be used. When working with lead, always keep the area well ventilated as lead fumes can be dangerous. While lead is poured at much lower temperatures than most metals, the same safety procedures apply.

WOMB FIGURE. Robert Lockhart. 1963. Cast lead. *Courtesy, artist*

►

THE RIVER. Aristide Maillol. 1943. Cast lead, 53¾ inches high, 7 feet 6 inches wide. *Collection, The Museum of Modern Art, New York*

PORTRAIT HEAD. Gaston Lachaise. 1928. Cast bronze, 17½ inches high.

Walker Art Center, Minneapolis

10

Metal Finishing

A PERFECT surface is rarely achieved in metal casting regardless of the efficiency of the mold process. Surfaces generally require cleaning up and retouching until the desired finish is reached. The core pins and other protrusions remaining from the casting process must be removed. The surface itself may have "fire scale," which is an oxide on the metal. Sometimes a metal surface will have cracks and other defects, such as air bubbles caught in the molten metal, all of which have to be repaired by filing or filling or both.

The protrusions remaining from the lost-wax process, sprues, vents, and risers, which are solid bronze, are removed with bolt clippers or a hacksaw, then a process called chasing is begun. This involves mechanically finishing or tooling a metal surface using a chasing hammer and other mechanical and hand tools shown on the following pages. Chasing tools and methods differ for individuals and for specific projects. Tools may be purchased ready-made or fashioned from tool steel. Ends can be shaped as a job requires. Striking ends can be round, flat, oval, circular, or rectangular; tool ends should be tempered.

The striking surfaces of chasing tools are generally polished smooth, but sometimes a slightly pitted surface will yield the necessary effect in the sculpture.

Following the chasing procedure, the metal surface may be further refined by filing and scraping, then sanding with wet/dry carborundum paper. A final polish with a good metal or buffing compound will produce a high finish if desired. The finish may be protected with a spray coat of clear acrylic lacquer or wax.

Sometimes, a special patina, or surface coloring, is desired for metal. A patina is achieved by chemically altering the metal surface, and many formulas for producing this controlled corrosion of the surface are given later in this chapter. It is rarely possible to achieve the same patina on two pieces of different metals even though the same formula is used. Factors that affect patination include the temperature of the chemical, the purity of the ingredients employed, the care exercised in preparing the formula, and the method of application.

In addition to applied patinas, metal surfaces can be changed by exposing the metallic object to chemical fumes or baths that will have a corrosive effect. The effect is arrested by dipping the metal in a stop-bath of water. One metal may be electroplated to another metal. Opaque lacquer or enamels may be applied where color is desired; or, if the original color needs protection, transparent coatings are applied.

Philip Grausman's approach to finishing cast metal pieces is based on training from Italian craftsmen steeped in the historical tradition of fine sculpture. He carefully and laboriously proceeds from one step to another, each time bringing the inherent qualities of the metal closer to the final surface finish. The sculpture *Pea* was originally conceived and executed directly in hammered copper sheet and was a finished work in itself, but it also served as a prototype for a bronze casting. A plaster support mold with a gelatin inner mold was taken from the copper and then a wax impression was taken from the flexible gelatin mold. The wax model was refined and the piece cast by the lost-wax method. The entire surface of the piece was then hammered, chased, polished, and protected with a spray coat of clear acrylic lacquer.

The piece *New Born* was originally modeled in clay and cast in stainless steel (Grausman sometimes uses nickel-silver, also called German silver, which has a color similar to stainless steel but the material is softer). The ceramic shell casting method was used and a considerable amount of chasing followed due to the hardness of stainless steel.

PEA. Philip Grausman. 1965. Bronze, 27 inches wide.

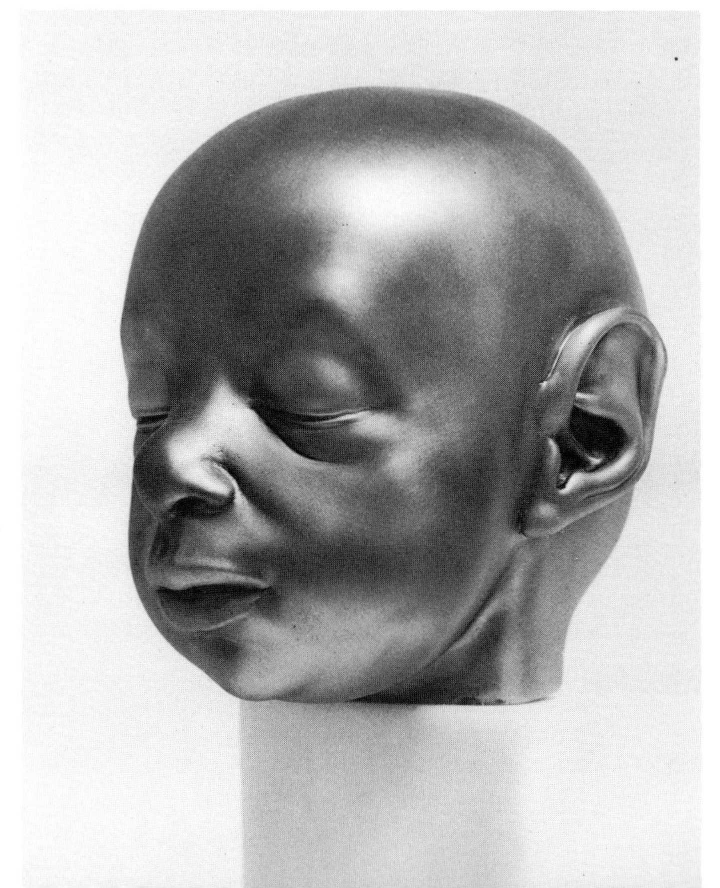

NEW BORN. Philip Grausman. Stainless steel.

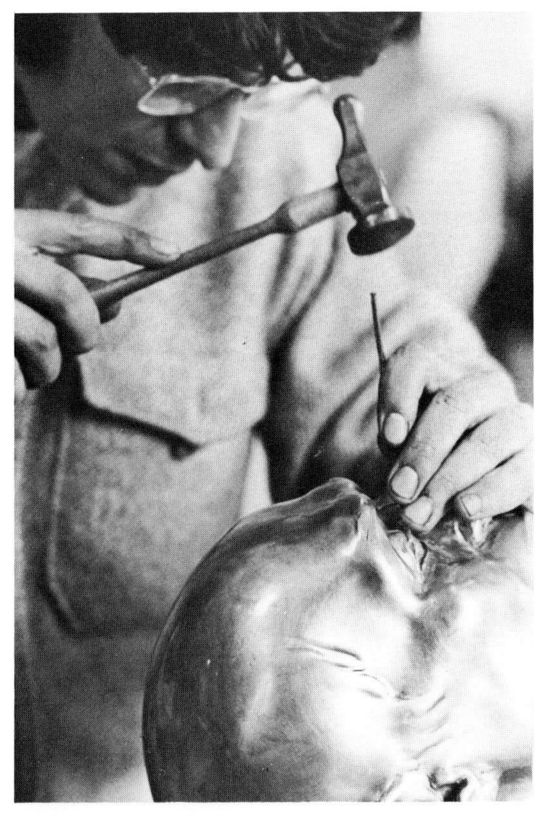

Philip Grausman uses fine chasing tools, files, scrapers, and wet/dry carborundum papers to achieve a high finish to form and surface.

HEAD OF JOSÉ LIMON—AMERICAN DANCER AND CHOREOGRAPHER. Philip Grausman.

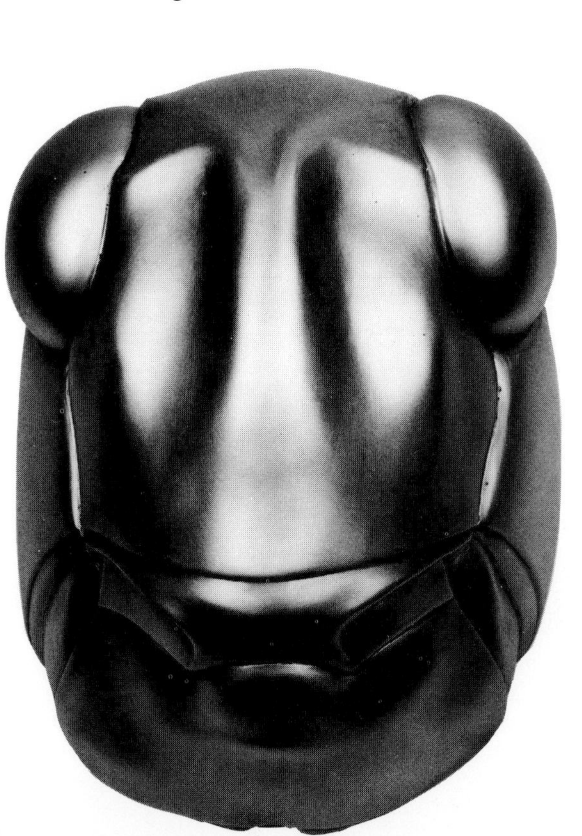

GRASSHOPPER MASK. Philip Grausman. 1968. Stainless steel, 9 inches high. This piece was partially polished and some areas burnished to a matte finish.

Vise and numerous tools used for the finishing processes: hammering, chasing, filing, and scraping. Sometimes, Mr. Grausman has the entire piece silver- or chrome-plated. Then the total surface of the sculpture must be hammered and chased to remove all porosity; if blemishes exist, plating does not cover them, so a homogeneous surface is essential.

Files, chasing hammer, and chasing tools with various shaped ends for achieving different effects on the metal surface.

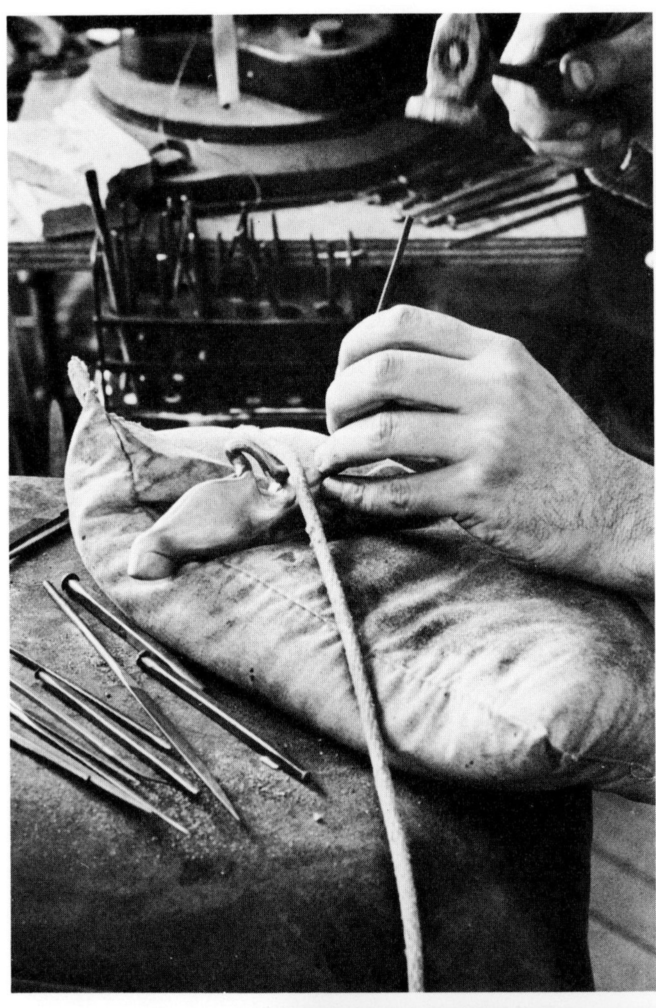

A small bronze piece is being chased. It is anchored to a leather sand bag by a loop of rope hanging down to the floor, which is held tightly over the piece with the foot while the work proceeds.

Fine finishing is accomplished with a small flexible shaft fitted with a disk backed with flexible foam. This setup enables you to follow delicate forms and transitions on the piece.

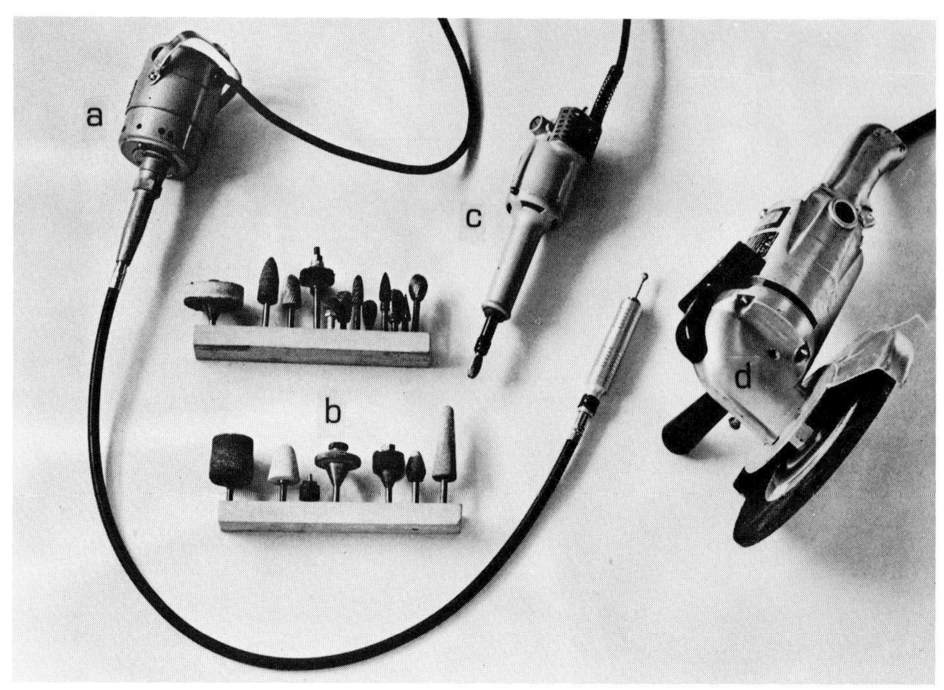

Left to right: (a) Small flexible shaft grinder fitted with a chuck; (b) an assortment of carbide rotary files and burrs, stone grinding wheels and shapes, sanding drums and shaped buffing or polishing wheels; (c) a high-speed (20,000 rpm) die grinder equipped with three collets, $\frac{1}{16}$, $\frac{1}{8}$, and $\frac{1}{4}$ inch; (d) a body grinder that can use large-diameter grinding wheels and disks at approximately 5,000 to 6,000 rpm.

An electric stationary belt sander with a thin strip of sanding cloth for sanding and polishing contoured pieces.

Two kinds of flexible shaft grinders (*above* and *at right*), both with variable speeds. Illustrated, with grinder (*right*), is a variety of attachments available: angle head and straight hand pieces, fiber-backed sanding disks, carborundum cut-off or grinding wheels, abrasive drum sander, and various polishing attachments.

A floor-mounted belt sander is equipped with a dust collector—especially useful when sanding plastics.

A spray booth can be made of wood framing and plastic. A wall-mounted exhaust fan keeps the air clear of floating particles that can be injurious to the health.

PATINA

The natural color of metal is often changed by the sculptor for esthetic reasons. In bronze, for example, color changes occur through oxidation over a period of time, but rapid oxidation can be accomplished by using various techniques that involve heat and chemicals.

The general procedure for establishing a patina, or color change, on the surface of metal is to begin with a clean surface. A patina varies according to slag content, ingot or remelted metal, alloy, and weather. Patinas release toxic or corrosive fumes, so good ventilation is important. Metal surfaces are cleaned by one of several methods: Clean or pickle the surface with acid such as one part sulphuric acid added to one part water; by sanding, grinding, filing, wire brushing the surface, or sandblasting.

The piece can then be left at room temperature or heated before it is dipped into a chemical solution, or the solution may be sprayed or brushed on. The piece can be placed in a vacuum-tight container or airtight plastic bag with an open dish filled with the chemical or a combination of chemicals. Bronze can be heated red hot and allowed to cool and develop its own color; or do nothing and let the piece develop its own color. You can bury the piece in sawdust, sand, shredded paper, or wrap with a chemical-soaked rag. You can also coat metal with any of the numerous paints available: oil, acrylic, polymer, among others. Or simply spray the piece a few times a day with salt water for a soft green color; steel will rust rapidly; lead and aluminum turn white.

Application of chemicals may have to be repeated numerous times and may continue for a week. Most patinas take only a few moments to develop; the piece is then washed when the desired color is reached.

To preserve a patina, many protective coatings are available: microcrystalline, bees-, or carnauba wax; plastic sprays such as acrylic, epoxy, or lacquer. No patina can be completely stabilized, especially if it is exposed to the outdoors. Epoxy and acrylic are the best stabilizers now available.

Certain polished-surface metals are more resistant to oxidation than others: stainless steel, for example, compared to bronze. A bronze surface, once polished, can be continuously reshined or coated with wax or a finish such as acrylic lacquer, but a thick coating will always show and is subject to scratching.

The final color of a metal casting can also be controlled by the alloy used or by small amounts of metal such as aluminum, silver, tin, and gold added to the crucible just before pouring. Temperature can also vary color; by overheating, the metal takes on a darker color. If the piece has been investment cast, the kind of wax used may also affect the metal surface by leaving a residue behind on the mold walls as the wax melts out.

Electroplating for bronze, steel, and aluminum and anodizing for aluminum, which is a protective coating and colorant, are permanent finishes.

Areas of a piece that are welded, joined, or filled in any way will usually patinate at a different rate or with a different color from the surrounding metal. For an even overall patina, try to use the same metal that was cast or cast rods for welding at the time the piece was cast.

Readily Available Chemicals for Patinating Bronze

1. Liver of sulphur and water
2. Ammonium hydroxide and water
3. Hydrochloric acid and water
4. Vinegar
5. Salt water
6. Ammonia
7. Raw linseed oil

Items 1, 2, and 3 are brushed or sprayed repeatedly on a hot or cold casting. Items 4, 5, 6, and 7 are brushed or sprayed repeatedly on a cold casting. Items 1 through 6 are applied using saturated rags, sawdust, or paper wrapped around cold casting.

WHEEL MAN. Ernest Trova. 1965. Silicon bronze. *Collection, Walker Art Center, Minneapolis*

Examples of Formulas for Patinas on Various Metals

Antique green for copper: 2 ounces of salt and ½ gallon of vinegar; immersion of the piece is necessary till color is reached.

Black for copper or golden brown on bronze: 1 ounce of ammonium sulphide and a gallon of water; piece may be heated or immersed.

Green for brass: 2 ounces of sal ammoniac, 4 ounces washing ammonia, and a gallon of water, immerse the piece.

Black on bronze: liver of sulphur or sulfurated potash in lump form that should be dissolved in water and brushed on a hot piece, or the piece can be wrapped.

For darkening aluminum: preheat. Carbon black from acetylene flame, brush in silicone wax; heat again and rebrush. Drano, a commercial drain cleaner, or lye will also darken aluminum.

Blue

Sodium hyposulphite (or) sodium thiosulphate	60 grams
Nitric acid	4 grams
Water	1 quart

Application: Dip.

Yellow Green

Sodium thiosulphate	1 ounce
Iron nitrate (or) ferric nitrate	8 ounces
Water	128 ounces

Application: Brush. After color is developed dip piece for approximately 5 minutes in diluted nitric acid; wash and dry as usual.

Blue

Ammonium hydroxide liquid placed in a chamber will produce ammonia gas as it evaporates which combines with copper ions for a deep blue.

Black

Copper carbonate	2 ounces
Ammonium carbonate	4 ounces
Sodium carbonate	1 ounce
Water	32 ounces

Application: Boil solution and dip.

Green

Cupric nitrate	1 teaspoon
Water	1 pint

Application: Heat and brush.

Antique White

Bismuth nitrate	2 teaspoons
Water	½ pint

Application: Heat and brush.

Brown to Black

Ammonium sulphide	1 teaspoon
Water	1 pint

Application: Heat and brush.

Brown to Black

Potassium sulphide	a few crystals
Water	1 pint

Application: Heat and apply.

Brown to Black

Antimony sulphide	2 ounces
Sodium hydroxide	4 ounces
Water	255 ounces

Application: Mix by weight; heat and apply.

Black for Iron

Apply aluminum sulphate full strength with brush. Clean rust from steel and iron with *phosphoric acid*. Rust steel or iron with salt water. Fire scale must first be removed.

Lead

Clean metal. Brush surface with 10 parts water to 1 part nitric acid. Fumes are toxic. Wash surface with water when desired color is reached. Protect with linseed oil immediately as patina is fragile.

BOUQUET. Robert Thomas. 1966. Painted bronze, 12 inches high.
Courtesy, Adele Bednarz Galleries, Los Angeles

Most. Bradford Graves. Cast concrete. (See demonstration series page 202.)

11

Concrete and Plaster Casting

CONCRETE is cement (calcined lime) mixed with an aggregate such as sand, gravel, or stone. It is the same material used commercially for buildings, sidewalks, roadwork, bridges, and other industrial applications. Concrete sculpture, therefore, is particularly adaptable to outdoor sites because of its imperviousness to a multitude of weather conditions and for its strength. Unlike metal casting, concrete does not depend on heat to gain fluidity; fluidity depends on water mixed with the cement and aggregate, which are semifluid even in their dry state.

Concrete, and its various mixtures, is often referred to as cast stone. The surface can reproduce an original very accurately. The smoother the form or mold surface, the smoother will be the concrete surface.

For large architectural pieces, forms are wood, steel, or plastic. For smaller pieces, any of the molding materials already discussed may be used. With wood forms, the inner surface must be sprayed with a commercial form oil (such as linseed oil) to act as a mold release and hold moisture in the mold for a dense surface. Steel forms tend to produce a tighter surface. Plastics such as sheet polyethylene produce a hard, highly polished finish, but there is a problem with wrinkles in the surface. Plaster of paris is also used as a forming material for concrete sculptures. At the Vermont Concrete Symposium (discussed in this chapter) earth, itself, was also used as a form.

When casting concrete, it may be mixed in a conventional concrete mixer, or it may be mixed by hand in a wheelbarrow and shoveled into the form. Molds may be packed with dry concrete mix and water slowly poured in over the mix. The concrete will accept only the amount of water needed to set it, which results in a harder than usual casting.

Castings should be allowed to remain in the mold for as long as possible; the longer concrete stays wet the stronger it becomes. Oiling or waxing a plaster or wood mold seals the mold so moisture cannot escape, while also serving as a release agent. Curing stops when the concrete dries out.

There are many possible mixtures of cement aggregates that will create different surface textures, color, and porosity of the finished material. The name *neat cement* is applied to cement mixed with water only. *Mortar* results when only a very fine aggregate, such as sand or marble dust, is mixed with the cement. *Concrete,* as mentioned earlier, results when the cement is used as a binder for aggregates such as sand, marble chips, aluminum oxide, gravel, crushed terracotta, or grog.

Portland cement (a variety and not a trade name) is widely used for sculpture. It is a fine gray or white powder with a setting time of about 3 hours. When it is mixed with water it binds the aggregate material together. As it hardens, it gains strength even greater than that of the aggregates mixed with it. When a lightweight aggregate such as Perlite (the trade name of a material derived from volcanic rock) is mixed with portland cement, the result is a lightweight casting of good strength and durability.

Plasters and their use in making molds and for casting are discussed in the section on plaster beginning on page 224.

Bradford Graves describes the procedures for creating the sculpture *Most* on page 200.

A wooden armature was built with 4 × 4, 2 × 4, and 1 × ½ lumber. The main supports are the 4 × 4s bolted to the platform. Shown is the armature for the larger of the two sections seen in the finished sculpture. Rope was tied between the wooden boards to hold the clay. This also binds the wood together to support the weight of the clay. The clay was squeezed around the rope and wedged between the wood. The clay was kept wet by covering it with plastic when not being worked.

Clay was applied over the armature starting from the ground up so that the clay would help support its own weight. The finished area was covered with wet burlap and plastic sheeting to keep the clay from drying out. In this photograph, one section of the wooden armature is completely covered with clay. The first layer roughs out the form. This was done completely with large chunks of clay. The clay was then ready to be modeled for specific texture and forms. The bottom section of the piece is the second layer now worked with smaller chunks of clay. The wood projecting out of the first layer helps to hold the clay's weight.

After work on the clay was completed, 59 plaster piece molds were taken from the two sections. Wooden braces were plastered onto each section to guard against the plaster warping and to make it easier to remove each plaster section. The plaster is very heavy, which is why so many sections are necessary in a sculpture of this size. Each section of plaster is separated by strips of aluminum; this prevents interlocking of the plaster sections and makes it easier to remove each section. The plaster sections were then removed from the clay and all clay removed from the mold's surface. The mold was dried of all moisture and coated with shellac, then waxed. The shellac and wax act to seal the mold's surface; they prevent the concrete from sticking to the mold and trap the moisture in the concrete for a four-day or longer drying period.

The various plaster mold pieces were aligned and tied together securely. The plaster mold was braced from the outside with a wooden structure to support the weight of the concrete pushing out from within the mold. The concrete was poured 10-inch layer upon 10-inch layer and was hand packed. The piece, reinforced by iron rods, rests upon a concrete foundation 3 feet thick. It was sunk below the ground's surface so that grass would grow between the forms.

After curing for four days, the plaster mold was removed and surface malformations due to warping or slipping of the mold were removed and sections carved away. The surface was machine sanded and a sealer was put on the surface to prolong the curing for the finished concrete sculpture MOST by Bradford Graves.

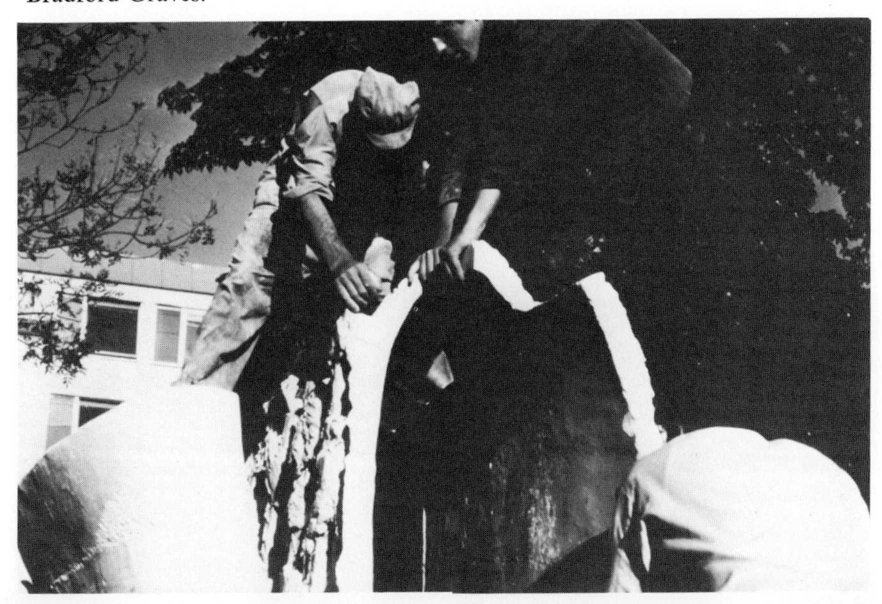

The works illustrated on these pages are by sculptor Rolando Lopez Dirube from Puerto Rico. Photos, courtesy artist.

Above: Reinforced concrete façade in progress. The relief is for United Artists Eastern Theatres, Inc., San Juan, Puerto Rico. 1970. The four façades of the building are totally covered with relief castings that were made flat on the floor and then lifted with cranes and bolted to the wall. The white areas show the foam boards that were cut into, forming the negative space into which the concrete is poured.

Opposite, top: SCULPTURE. Reinforced concrete. Installed at the park of the La Arboledo development, San Juan. The three parts were cast separately and steel H beams used in the structure. Molds were made of wood. The piece was assembled with a crane.

Opposite, bottom: MURAL. Havana Riviera Hotel, Havana, Cuba. 1957. Reinforced concrete. The concrete wall relief was cast into molds made of wood, galvanized steel, plywood, and clay. For added texture, venetian gold mosaics were put into the molds and picked up in the cement as it was cast. Powdered coloring was added to the concrete: ocher, sienna, sevilla red earth, and other pigmentations. Colored glass cullets were also cast into the wall to serve as "windows" later.

The following series illustrates the procedures used by Gordon Woods for casting a concrete cornerstone. Photo series, courtesy artist.

Cornerstone in progress. It will be 114 inches high, 34 inches wide, and 30 inches deep. Mr. Woods usually develops such a piece with drawings and maquettes from which the full-scale piece is constructed.

The form lumber has been selected, cut, and assembled to achieve the desired texture or surface design. Styrofoam, cut to a rough shape, will be carved to achieve a final negative shape. A Styrofoam box (*at right*) forms the male element of the key that will lock the finished slab to the second element.

The reinforcing iron has been cut, bent, and welded together to assure that it will not move during casting.

The form has been closed and filled with concrete that has been vibrated by the "vibrator" shown in the foreground. The vibrator assures a dense cast without air bubbles and a close reproduction of the surface presented by the form lumber.

After a week to ten days of curing (it is wrapped with plastic in the form), the slab is raised for removal of the foam and the first rough carving. The foam is readily removed by sandblasting, and carving is accomplished with a pneumatic hammer and point chisel.

The second section is formed from re-sawn 1 × 4 pieces of lumber. Secondhand materials can always be employed satisfactorily and inexpensively for these forms. The key block shown on the photo on bottom of page 206 will now form the female element when the concrete is poured against it.

Reinforcing rods are added and blocks of foam are tied to it to lighten the cast. These will remain buried in the sculpture.

The panel that will appear inserted in the second section is prepared by assembling newspaper printing mats and pressing them into bread dough, which is then baked. The newspaper mat impressions become an integral part of the design and texture.

The panel as it appears after the dough has risen and been baked. It was then shellacked and a flexible latex mold was made from the bread. Concrete was then poured into the latex mold.

The form for the second section is assembled with the concrete panel set in place and ready for pouring. After removal of the form, the two elements are joined at the key. The protruding irons are given their final curvature. Additional carving completes the sculpture.

FIGURE III. Hans Aeschbacher. 1968. Cast concrete sculpture, 700 centimeters high, 350 centimeters wide, 35 centimeters deep. (See following series for development of this piece.)

Photos, courtesy artist

Detail of cast sections being erected.

Wooden forms for the finished sculpture FIGURE III (*opposite*) are set to define the shapes. Plastic sheeting is laid on the floor to assure separation. The inner sides of the wood have been coated with a release agent. Reinforcement rods are bent to the shapes and wired together.

Detail shows how reinforcement rods are wired together.

Concrete is placed in mold by hand. Workman at left is using a vibrator to settle the concrete and raise air bubbles to the surface. The surface is then troweled.

MONUMENT TO E. RUSJAN. Janez Lenassi. 1960. Cast concrete.
Courtesy, artist
►

FIGURE III. Hans Aeschbacher. Cast concrete fountain, 491 centimeters high, 290 centimeters wide, 240 centimeters deep. *Courtesy, artist*

In-process photos showing the development of the sculpture (*opposite*) by Janez Lenassi.

View into the upper part of the mold with steel reinforcement rods in place.

Preparatory work on the wooden molds.

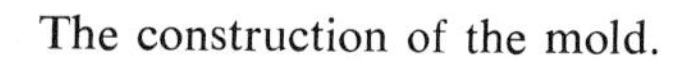
The construction of the mold.

A detail of the impression of the mold on the surface of the cement.

A MONUMENT TO THE DEAD PARTISANS. Yugoslavia. 1965. Janez Lenassi. White concrete, 9 meters high, 7½ meters wide, 7½ meters deep.

Photo series, courtesy artist

PROCESSION. Albert Vrana. 1966. Cast stone. Professional Arts Center, Miami, Florida. For his cast architectural commissions, Albert Vrana follows the basic procedures described in the following series on the opposite page.

Molds are made of Dyplast, an expanded polystyrene foam that is ideal for huge patterns that can be easily transported to any casting site. Since the mold is destroyed in the stripping operation after casting, no draft or taper is necessary in the mold.

Guidelines from large-scale drawings are transferred to the foam sheets, which are 4 feet wide in a 25-foot run. The large-scale forms are cut with an electric soldering gun, saw, or hot wire. A color code system, devised to aid assistants, determines the placement of the relief sections. These forms are glued on the foam sheets with Elmer's, Tite Bond glue, or contact cement. After the glue has dried, the shape is carved, ground, and textured to correspond to the design of the original model.

Release agents such as emulsified petroleum oil, paraffin, epoxy resin, or polyvinyl alcohol can be used to prevent concrete adhesion and change the texture of the foam if necessary. Stripping the molds from the casting before a total cure has taken place and scrubbing the surface with a wire brush allows the aggregate to show through.

The molds were transported to the Pre-Cast Corporation, Miami, Florida, for production of the cast panels. The completed panels were then transported to the building site.

Foam panels were assembled according to color coding.

Refining the design was partially accomplished by sanding.

The face coat was carefully cast to avoid air bubbles and to capture all details.

After the foam was stripped, the castings were cleaned with a mild acid solution commercially sold for cleaning concrete from tools.

STORY OF MAN. Albert Vrana. 1962. Cast stone. Miami Beach Public Library, Florida.

For the panels, above, instead of foam panels, Mr. Vrana used wet sand retained within dimensionally accurate wooden frames. The designs were modeled, in reverse, in the wet sand. Concrete was then poured over the sand and steel reinforcement bars were laid in the concrete, taking care not to allow them to protrude through the face of the design. After the concrete had set and cured, the castings were removed from the sand molds and cleaned.

Series, courtesy artist

VERMONT SYMPOSIUM

Several known and experimental concrete casting techniques were explored at the 1971 Vermont International Sculpture Symposium, Griswold Industrial Park, Williston, Vermont, under the direction of sculptor Paul Aschenbach. The accompanying photos illustrate some of the works accomplished.

Most of the sculpture poured at this symposium used a ⅜-inch mesh aggregate and a standard cement-to-aggregate proportion. Considerable handwork was done during the pouring, especially around detailed areas. A great deal of ramming with poles was also done during the pouring. In several situations an electric pencil vibrator was used in the concrete while in others the vibrator was applied to the outside of the form to help settle the mix and to eliminate air bubbles.

Forms must be watertight at the edges. If water leaks out during vibrating it will carry the local grout (cement) with it, leaving a "honeycomb" of exposed aggregate and concrete weak from drying too quickly. Rough edges and other problems can be filled in after the form is stripped off by adding a mix of fine sand and cement. The surface may also be honed down with a carborundum stone.

Sculptor Peter Ruddick's work includes a circular element that was poured flat and reinforced with steel bars.

A semispherical shape is carved from the earth and covered with cement and metal reinforcement wire by Paul Aschenbach. Organic forms are dug successfully in the earth while geometric ones are best cast in wood or metal forms.

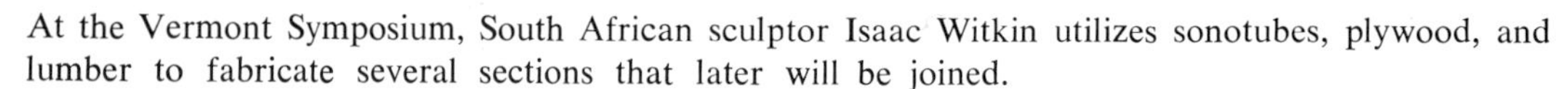

At the Vermont Symposium, South African sculptor Isaac Witkin utilizes sonotubes, plywood, and lumber to fabricate several sections that later will be joined.

Eduardo Ramirez from Colombia, South America, pours concrete with a boom pump. The outer form is held square with a steel column and clamps.

Plywood inserts are placed and weighted onto the surface of the slab while the concrete is still soft by sculptor Carl Floyd. The resultant shapes form the relief on the reverse side of the sculpture. Masonite was also used to make the form. The concrete was poured on a polyethylene sheet for a smooth surface.

Vermont Symposium photos, courtesy Paul Aschenbach

CAST CONCRETE. Carl Floyd. Vermont Symposium. Concrete, 40 feet long, 8 feet high, 8 inches thick. Weight, 10 tons.

Photo, June Aschenbach

SCULPTURE. Peter Ruddick. Vermont Symposium. Cast concrete, 8 feet high, 30 feet wide, 8 feet deep. Weight, 8 tons.

Photo, June Aschenbach

SCULPTURE. Paul Aschenbach. Vermont Symposium. Cast concrete, 4 feet high, 12 feet wide, 4 feet deep. Weight, 5 tons.

Photo, June Aschenbach

PLASTER MOLDS

The use of plaster for sculptural purposes can be traced back about 5,000 years. Original plasterwork, plaster molds, and castings date to approximately 2500 B.C. from Egypt, and then continue to be in the cultural remains of the Greek and Roman civilizations. Sculptures, architectural ornaments, and death masks were commonly made directly with plaster or from plaster molds.

Plaster is derived from gypsum, a mineral rock first found in the earth thousands of years ago. When gypsum is heated to approximately 350° F. it dehydrates, or calcinates, becoming an inert white powder that later can be rehardened by mixing with water to produce plaster.

The setting time of plaster can be varied by the use of commercial additives called retarders or accelerators. For coloring, water-soluble pigments mix well with it. Plaster is inexpensive and accessible. It can be worked easily in all its stages while setting, and the expansion problems are predictable. There are several types of plaster used for sculpture. Plaster of paris has a comparatively rapid setting time of about 5 to 10 minutes. Plaster such as hydrocal, which is used for casting, is white, hard, and fine-textured. Molding plaster, with a setting time of 5 to 10 minutes, is stronger than plaster of paris but it is softer and easier to carve than casting plasters. Molding plaster is the most widely used of the various plasters.

To mix plaster, put about 1 to 2 inches of room-temperature water in a low pan or a plastic bowl. Sprinkle plaster over the entire water surface gradually until small

AXE I. Bradford Graves. 1967. Plaster with watercolor, 11½ inches high.
Courtesy, artist

mounds form over the surface; then stop. Allow the water to saturate the plaster. After a minute or two, mix by hand or with an electric drill. Use a mixing paddle for large amounts.

Never add more water and plaster to a mix; always mix a new batch when more is required. Do not use water that has even the smallest amount of used plaster in it. If you use water that is too hot, the plaster will tend to set too rapidly. Too much water will make the mix thin and runny. The ideal mixture for plaster is about the consistency of sour cream.

CASTING FROM CLAY INTO PLASTER

Demonstration by Peter Ruddick.

A two-piece plaster waste mold is made from a clay model. The clay has been modeled over an armature made of pipe with metal loops and fastened onto the baseboard. Pieces of foam, small boxes, and wooden sticks can be used to build large forms as part of the internal armature. One-and-one-half-inch brass shims are placed along the high points of the mold, bisecting it. The piece is propped on one side with used sticks temporarily.

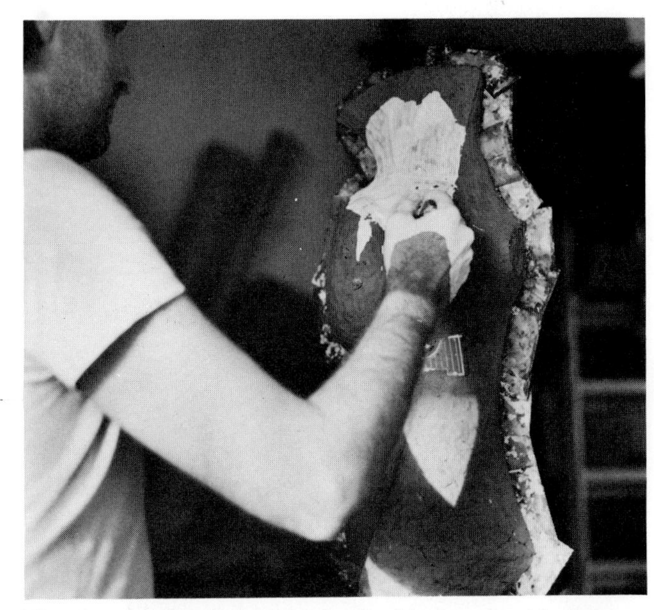

The first plaster layer is mixed with laundry blueing so that after the piece is cast and the mold is being chipped away, you can differentiate the casting from the mold.

The blue plaster is splashed directly onto the clay—do not brush on—and the buildup with white plaster continued with a spatula. A handful of plaster is scooped from one hand with a spatula and placed onto the mold. It can be pushed about with the spatula long after it has begun to set. Build up the plaster to about a ¾-inch-thick wall. Always try to keep the mold as light as possible as it must be later chipped away, which could damage the casting. After the first half is completed, the wood support on the other side is removed and the second half of the plaster waste mold is made, first applying a layer of blue and then building up with the white.

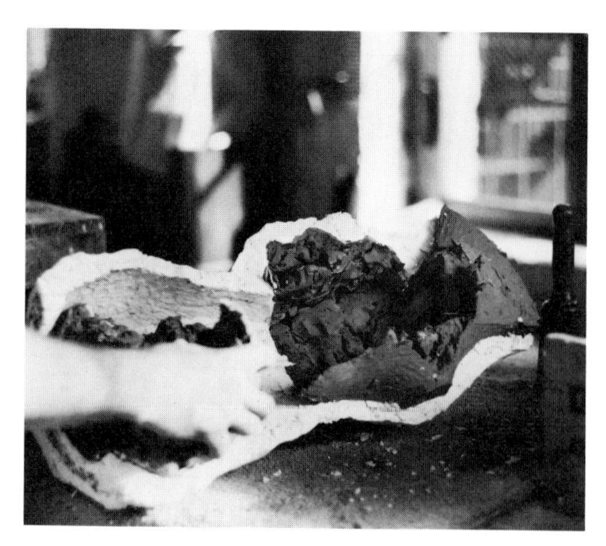

After the plaster sets, the shims are pulled out and the two halves of the mold are separated at the seam and the clay is removed. If suction between the sections is too great, gently drive wood wedges and run water between the sections. Any small bits of clay that stick to the mold can be removed by balling up some clay and rubbing the inside of the mold or by washing out the mold with a hose and soft brush.

Brush a coat of shellac over the entire inside of mold to help separate the mold from the cast piece later on. Apply a coat of paste wax, green soap, or commercial mold release to entire casting area, seams, and bottom of mold.

Two halves of the mold are reassembled. They may be tied or wired together. A layer of plaster, clay, or wax is placed over seams to prevent the plaster from leaking when it is poured into the mold.

After the mold is dry, the inside coated with shellac and separator, the sections wired together and the seams sealed, a batch of plaster is mixed and poured into the mold. Here, Mr. Graves turns the mold, forcing the plaster to coat the entire negative space. He will do this several times building the plaster up to a thickness of approximately 1 inch and then reinforcing it with burlap.

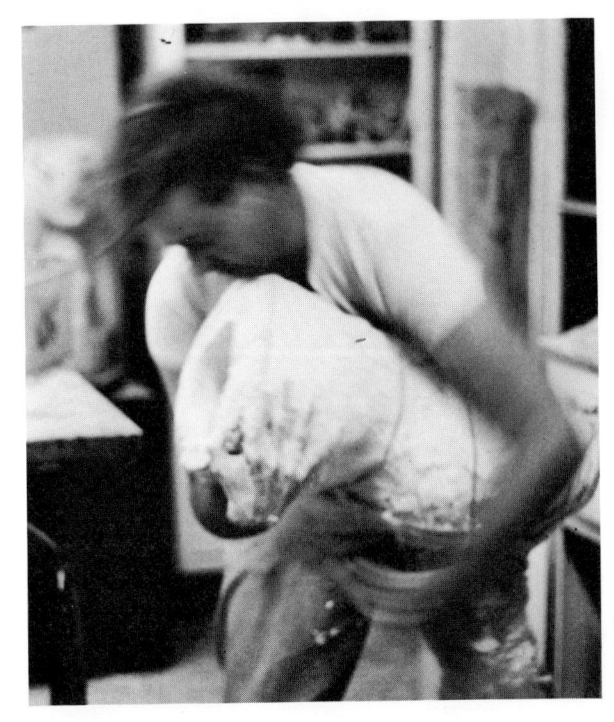

The hollow plaster casting is seen from the bottom, with most of the plaster piece mold chipped away. The seam can be seen running along the top. In some cases the casting can be poured solid with one batch—much depends upon the size and complexity of the casting. The casting can now be refined: dried, left white, or colored.

SUSPENDED DISC #5. James Wines. 1966. Painted cement and steel, 88 inches high, 60 inches wide, 18 inches deep.
Marlborough-Gerson Gallery, Inc., New York

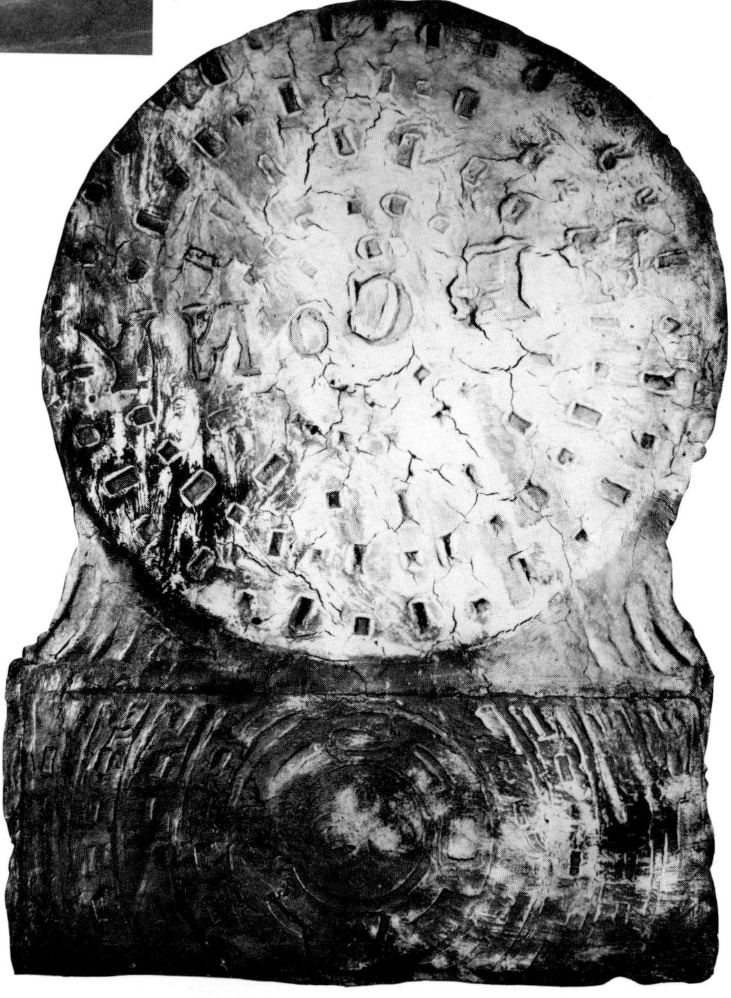

GANSEVOORT STREET. Gillian Jagger. Plaster and tar, 46 inches high, 36 inches wide.
Ruth White Gallery, New York

RETROSPECT. David Schneider. 1968. Cast cement, 18 inches wide.

Courtesy, artist

STUDY FOR THE PAVILLON DE FLORE. Jean Baptiste Carpeaux. 1865. Plaster.

The Art Institute of Chicago

The pieces were made originally in plaster for either bronze or white epoxy casting. Steel reinforcement was positioned and bent according to the shape to be cast. Then glass and plastic sheeting, which are self-releasing from plaster; weights were used to hold sheets in place during plaster pour which was leveled by gravity and a final scraping. After casting, the plaster was worked with files, rasps, and other finishing tools. Then a polysulphide rubber (Black-Tufy) or silicone rubber mold can be made from the plaster and an epoxy casting taken from that. Direct sand casting can be made from the plaster also. Since the bronzes are polished, a good surface is necessary; therefore he prefers French sand casting that produces a homogeneous surface.

UNTITLED. Peter Ruddick. 1970. Cast plaster, approximately 42 inches tall.

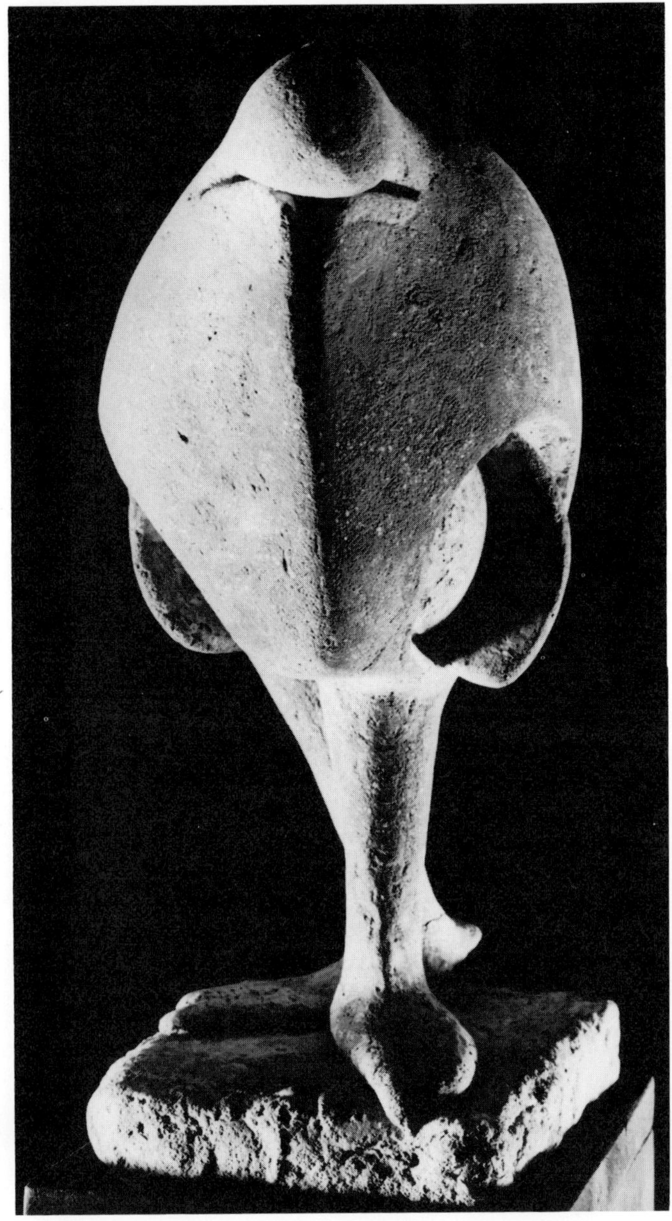

BIRD. Robert Lockhart. 1964. Cast cement, 12 inches high.

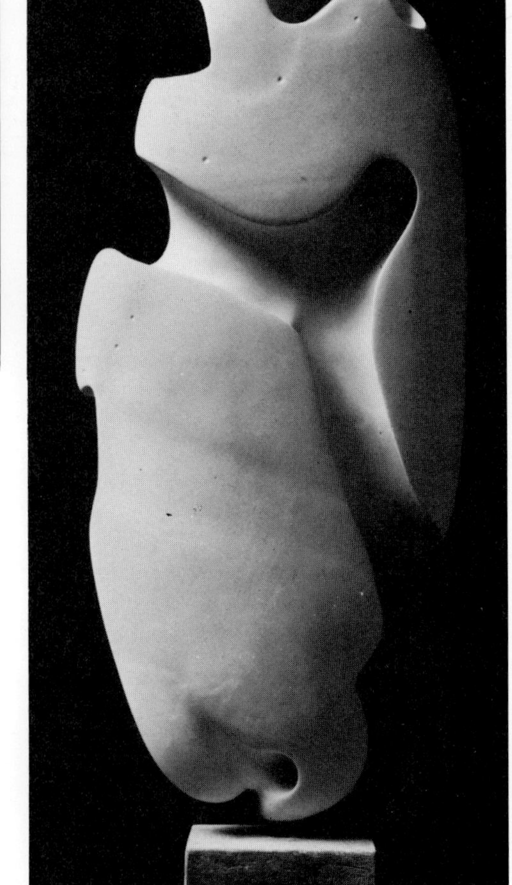

UNTITLED. Oddo Aliventi. Cast plaster, approximately 20 inches high.

Courtesy, artist

CHAIR. Ruth Francken. 1971. Plaster model of chair to be cast into aluminum.

Photo, Jack Nisberg

STANDING FIGURE. George Griffin. 1969. Cast concrete.

Courtesy, artist

CLAUS SLUTHER, SCULPTOR. Henry Bouchard. Plaster, 7½ feet high.

The Art Institute of Chicago

WALL FOR PALAZZO INADEL, ITALY. Oddo Aliventi. Cast concrete. *Courtesy, artist*

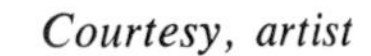

SCULPTURED WALL. H. Carlo Ramous. 1963. Cast cement. *Courtesy, artist*

12

A Student-Built Foundry

As casting techniques continue to be taught and developed at several learning levels, it has become evident that young people, with proper supervision, can carry through the necessary building and preparations for a foundry and kiln. These are valuable steps in rounding out a complete knowledge for future sculptors and teachers who wish to be totally familiar and involved with the technical processes. It should be emphasized that knowledgeable supervision is essential, yet students should be given enough latitude to think through and develop the construction from plans.

The following photographs illustrate the building of a below-ground foundry by fifteen-to nineteen-year-old students at the Hinckley-Haystack School of Arts and Crafts in Maine under the instruction of George Greenamyer of the Massachusetts College of Art, Boston.

The detailed drawing of the foundry installation designed by William Enright on the following pages may be used as a basis and adapted to individual space and needs. The following notes for the construction of a below-ground furnace should be carefully consulted as they apply to the drawings on pages 236 and 237.

Notes on the Construction of a Below-Ground Furnace

All dimensions are variable according to individual needs; the inside diameter of the brick liner should be 4 to 6 inches larger than the largest crucible to be used, the space necessary for the crucible tongs used to lift the crucible from the furnace. Capacity as designed: #150 crucible.

Standard concrete mix for walls and floor—forms made from 2 x 4s and untempered ⅛-inch Masonite.

Approximately 400 insulation bricks, 2600°, and 130 fire bricks, 3300°, were used.

Burners were two "Maxon Premium," premix burner M-250, Muncie, Indiana, for propane, 20 p.s.i.; to start fire reduce pressure to 12–14 p.s.i.

Pyrometer—Newton Potters Supply, Inc., Chesterfield, Ohio.

Silicone carbide lid, crucible, and crucible rest from S. C. Coler, Bay State Crucible Company.

Standard commercial crucible tongs, shanks, skimmer, ingot molds, and chain pulley system mounted on an overhead I-beam were used. Design, William Enright.

A BELOW-GROUND FURNACE

Dig a shaped hole in the ground 3 feet deep and square the walls.

Build a form from ½-inch plywood, ¼-inch untempered Masonite, and 2 × 4s as indicated in the drawings on pages 236–237.

Drawings for the construction of an underground furnace.

Design, William Enright
Drawings, Lebbeus Woods

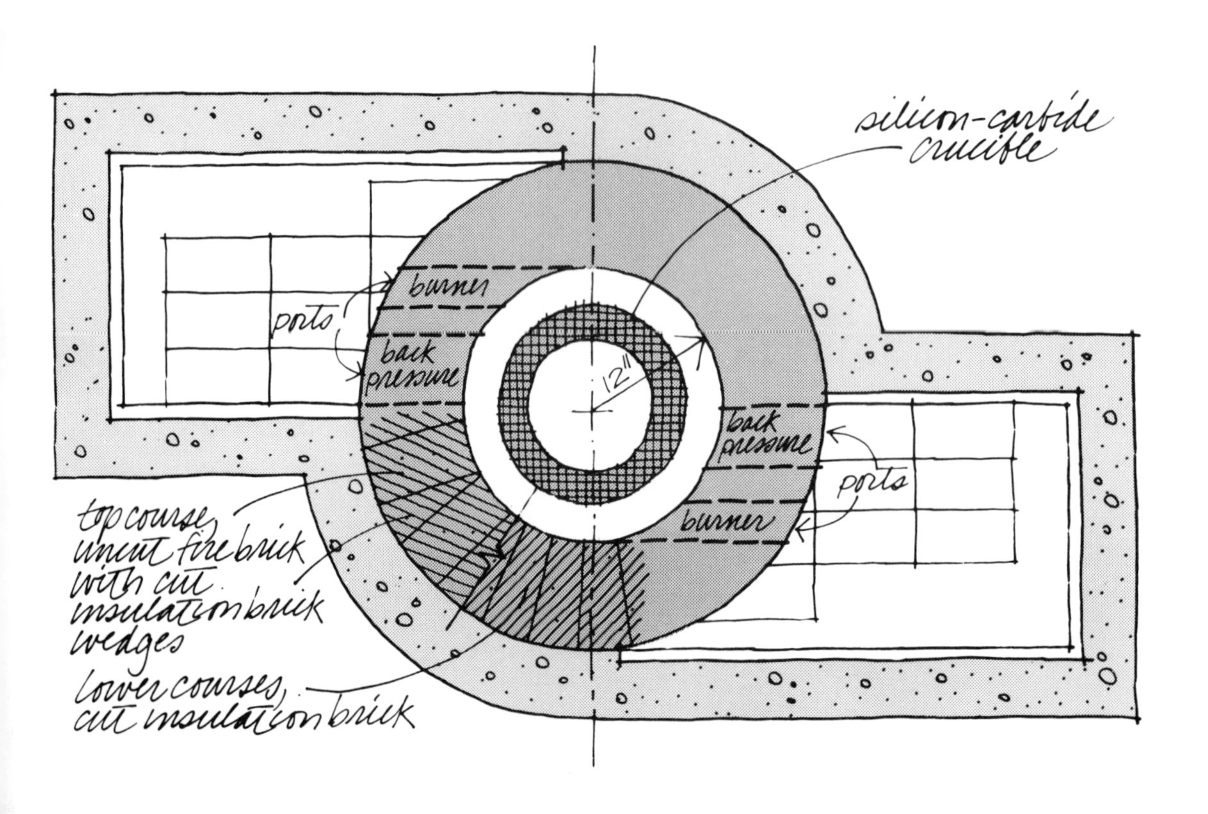

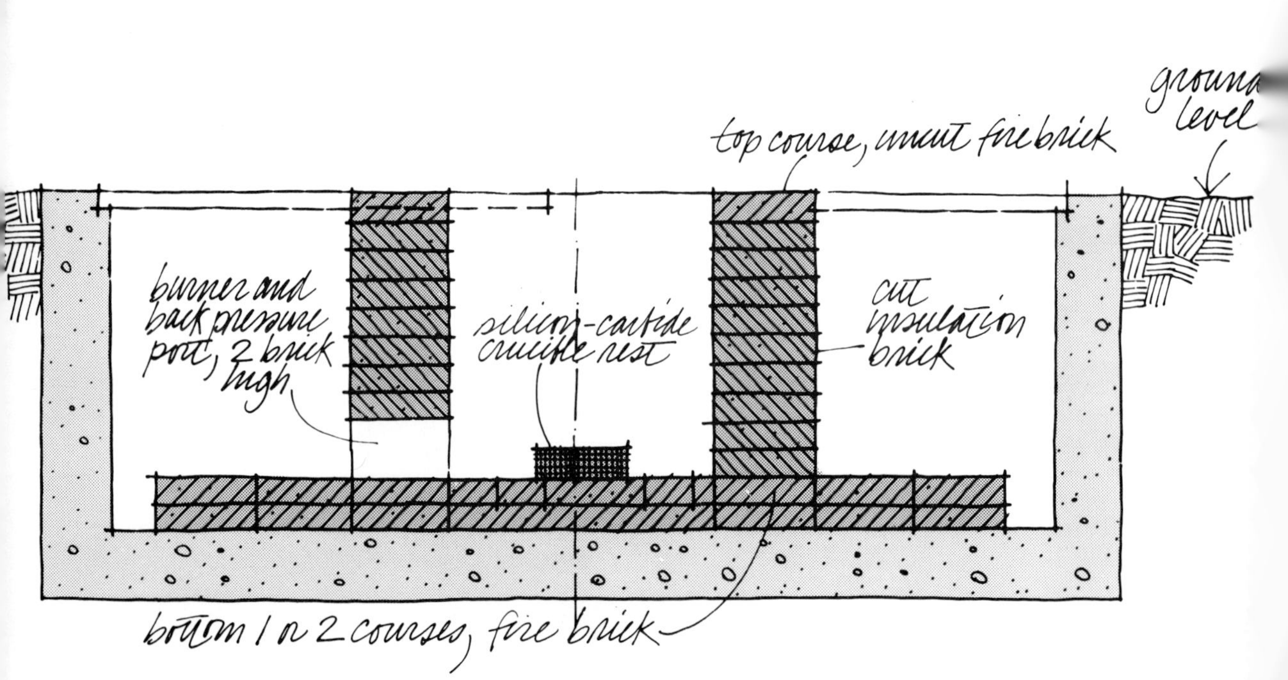

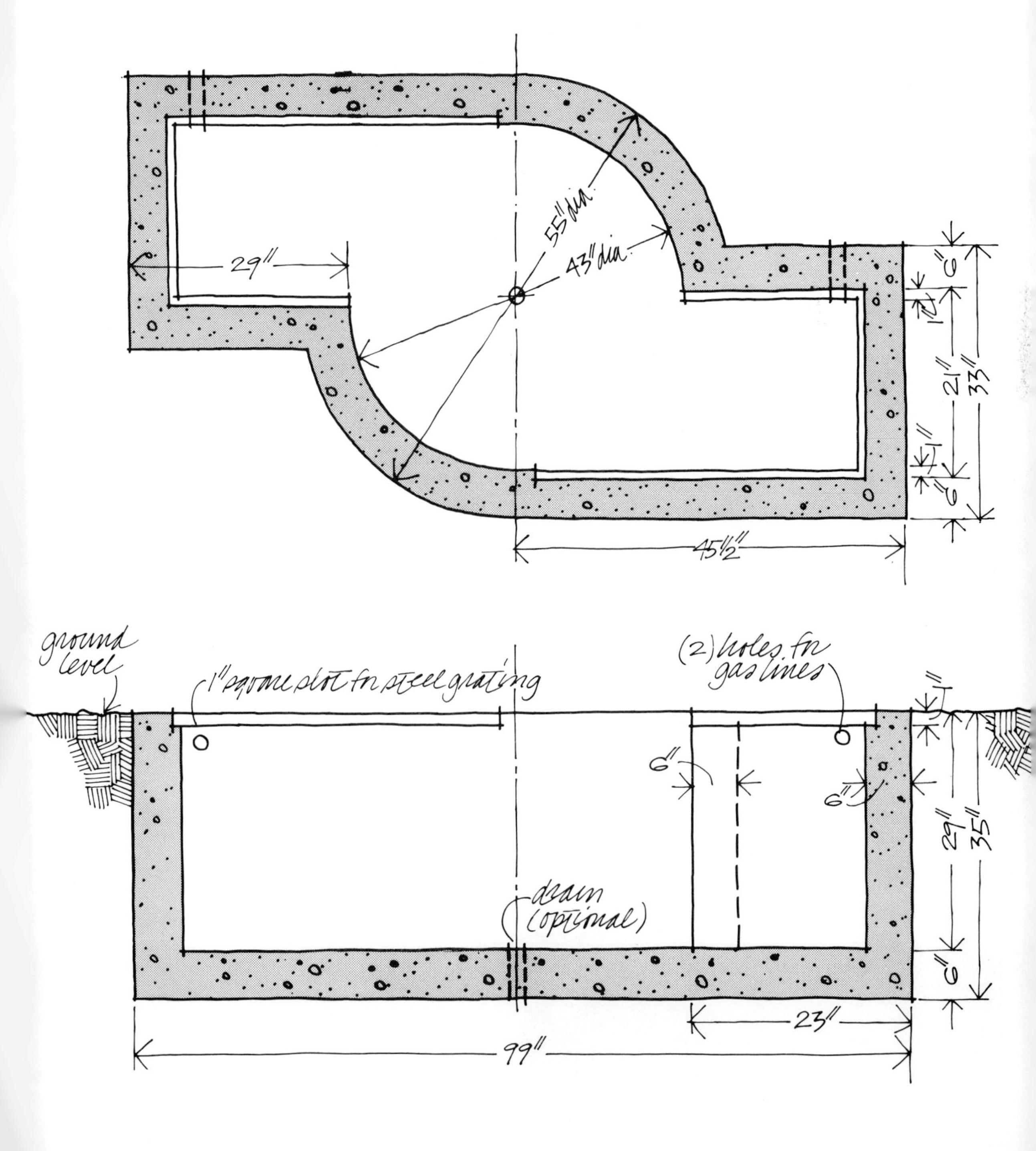

55" dia.
43" dia.
29"
6"
1"
21"
33"
45½"
ground level
1" square slot for steel grating
(2) holes for gas lines
6"
6"
1"
29"
35"
drain (optional)
23"
99"

The concrete is poured (approximately 6 inches thick throughout) with an optional drain formed by a metal can set in the floor. Pour walls immediately, or the floor can be allowed to set before the walls are poured, in which case reinforcement rods extending from the floor should be used. If the form is put in place on soft concrete the form will have a tendency to float and must be weighted down. Two metal cans are set in opposite walls of the pit in the unset concrete walls where the gas lines will enter and be attached to the burners.

A sand pit with concrete walls is also dug and a form made, which, when stripped away, will have slots in the walls to receive dividers for different-sized tamping areas.

After the concrete is poured, the surfaces are troweled smooth and covered with plastic sheeting for a few days to keep the concrete damp while it is curing. After curing, the wooden forms are stripped away. Note placement of metal cans for gas line.

The finished below-ground furnace is shown with a silicon carbide crucible and lid. The metal frame with loops is for lifting lid with an overhead hoist. Steel grates are cut to fit openings over the gas lines and were bought at a metal junkyard.

To build the brick interior, two courses of firebrick are laid for the bed of the furnace, then insulation bricks are cut to fit the circular shape. The very top layer is firebrick, used for its refractory qualities and strength. Two holes are located at each end in the first course of insulation bricks; one for each burner, Maxon Premium burners M-250, and the other two for back pressure, which is adjusted with insulation brick as necessary. Firebrick was used for a crucible rest, but a silicon carbide one may be bought.

Detail of furnace with burners lit. One burner flame is seen at right, the other enters from the opposite side. Between them they create a strong swirling action that travels around and up the walls of the crucible and out through the center of the lid. Detail shows how the top row of firebrick is laid over the furnace walls of insulation brick.

Three views of the burners used in the furnace; they can be used for the burnout kiln also.

Gauges used throughout for propane gas pressure control on furnace and burnout kiln. The storage tank gas pressure was reduced by larger gauges before entering the smaller gauges.

SAND CASTING FROM FOAM ORIGINALS

Students are riddling, or sifting, plain river sand through ¼-inch, then ⅛-inch hardware cloth. The sand is then mixed with approximately 10 to 20 percent Gordon clay (or if bentonite clay is available, the 10 to 20 percent can be made with ½ Gordon and ½ bentonite) and a small quantity of water, about 5 to 10 percent, in the portable cement mixer shown at right. When the clay-sand and water mix just retains the shape of your hand after you have squeezed a handful you have added enough water; too much water creates steam that causes a poor casting.

A low-density foam is used by the students to make patterns to be cast in aluminum or bronze. The foam is cut on a bandsaw or with a hot wire and shaped with knives and files. Pieces are glued together with a white glue or contact cement. After the pieces are finished, a sprue, approximately one inch square, is attached to the piece, and a large foam pouring cup is attached to the sprue. Vents and runners are also added. Sand is more permeable than investment molds so fewer vents are needed for sand. Finished foam patterns are shown being packed, or rammed, with sand into metal flasks. Firebricks are used as a filler.

WAX ORIGINALS FOR INVESTMENT MOLD CASTING

Foam is again used in the lost-wax process; the sprue system is composed of low-density foam that provides a light, strong structure for the pattern. The foam will burn out of the mold along with the wax and moisture. The investment used is a mixture of ⅔ washed and riddled river sand and ⅓ molding plaster (gauging plaster can be used but it takes longer to set).

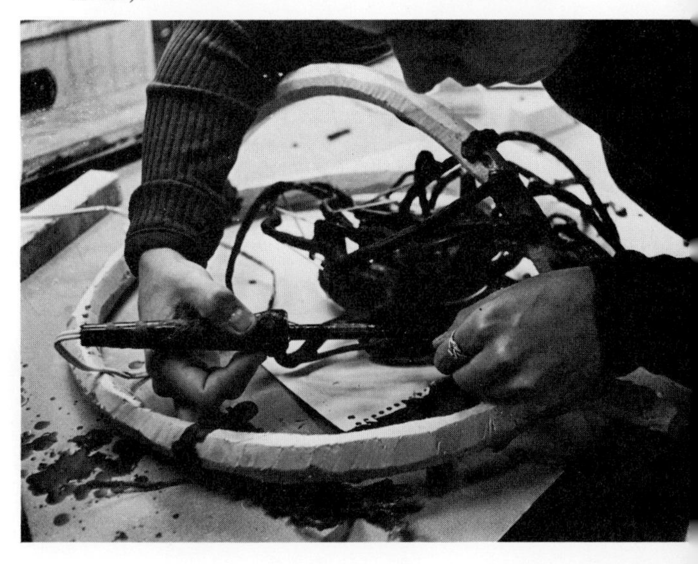

BUILDING A BURNOUT KILN FOR INVESTMENT MOLDS

When investment molds are completed, they are stacked in a kiln on firebrick. The base of the kiln is made with one layer of concrete block followed by two layers of firebrick for the floor. After stacking, all loose investment and other particles are vacuumed, as some molds have a bottom drain which means the pouring is up and foreign matter could fall into the mold.

The burnout kiln is shown completed and operating. For the walls two courses of firebrick have been laid from the floor of the kiln up. Two burners are placed, as shown, one on each side opposing each other; the openings must be left in the first course of bricks when the kiln is being built. A pyrometer will be placed at an imaginary center point that is marked by the light-colored insulation brick. The roof of the kiln is covered with two courses of insulation brick laid on rows of angle irons placed back to back and the brick in between. The burners, each with its own gauge, are lit and run at such a low pressure that it does not register on the gauge. Temperature is held at approximately 100° to 200°F for the first day and run up gradually and slowly to 1,000°F where it is held for a day, and then let down to approximately 500°F. The kiln is opened, the molds packed in the sand pit, and the metal is poured.

ADDITIONAL EXAMPLES OF HANDMADE FURNACES

When a temporary furnace was needed, one was built quickly on the site with firebricks and fired with a propane gas venturi-controlled burner to melt aluminum.

This furnace is made from a 50-gallon steel drum lined with firebrick and fired with propane. A Maxon Premix burner was used with a low-pressure blower that required another hole in the furnace wall for back pressure. If a powerful blower is used, no back pressure hole is necessary.

Example of a handmade venturi burner system created from black pipe, brass valve and lines, and a brick regulating air flow. Propane was the fuel.

CASTING METAL

Crucible being removed from furnace with an assist from overhead pulley system on a rail made from a used I-beam. Standard commercial crucible tongs have been modified with black metal parts for use with heavy loads and pulleys.

The pour is started. The shank has no safety device, so a third student must hold the crucible in the shank and a fourth student skims as the metal is poured. This clearly illustrates the amount of fumes that result from the molten metal vaporizing the foam pattern. It should be noted that when pouring investment molds no fumes from wax or water should be apparent—if they are, the burnout was not completed properly. There is no burnout involved when doing sand castings with foam originals; the metal is poured directly on the foam.

ANOTHER HAND-BUILT FURNACE

Prewarmed ingots are lowered into the hot crucible with tongs. For melting ferrous and nonferrous metals, clay graphite ceramic-bonded or silicon carbide carbon-bonded crucibles may be used. The clay graphite crucible is less expensive but the silicon carbide is stronger and conducts heat more quickly. No matter what kind of crucible is used, only one kind of metal should be melted in each one. New crucibles should be preheated or tempered before use and be as large as possible for the furnace used.

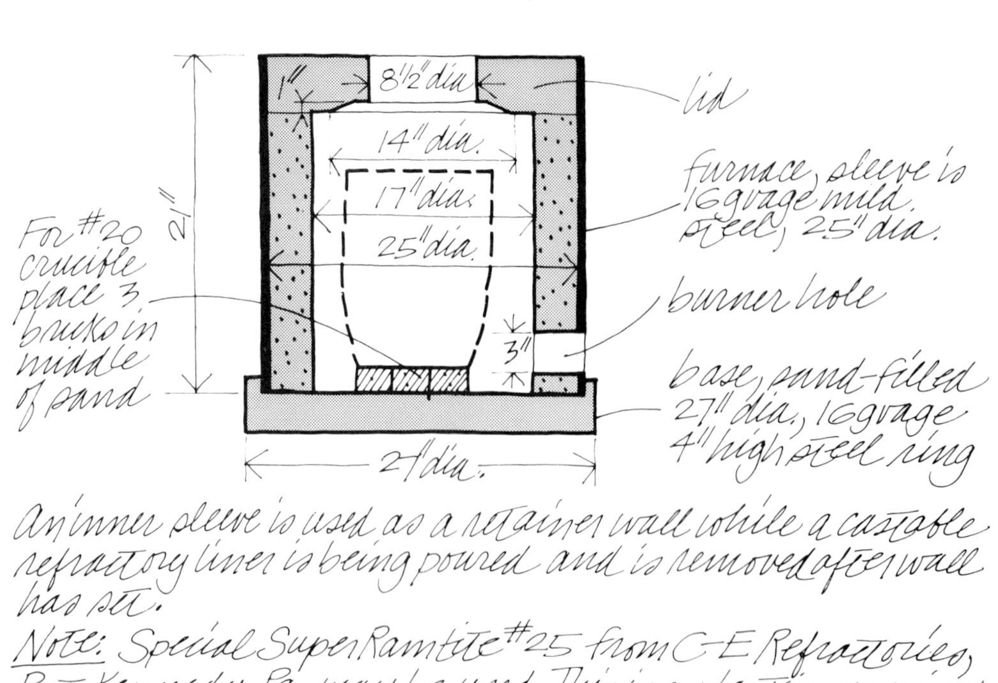

An inner sleeve is used as a retainer wall while a castable refractory liner is being poured and is removed after wall has set.

Note: Special Super Ramtite #25 from C-E Refractories, Port Kennedy, Pa. may be used. This is a plastic pre-mixed refractory and is rammed without a form into a steel sleeve and air hardens.

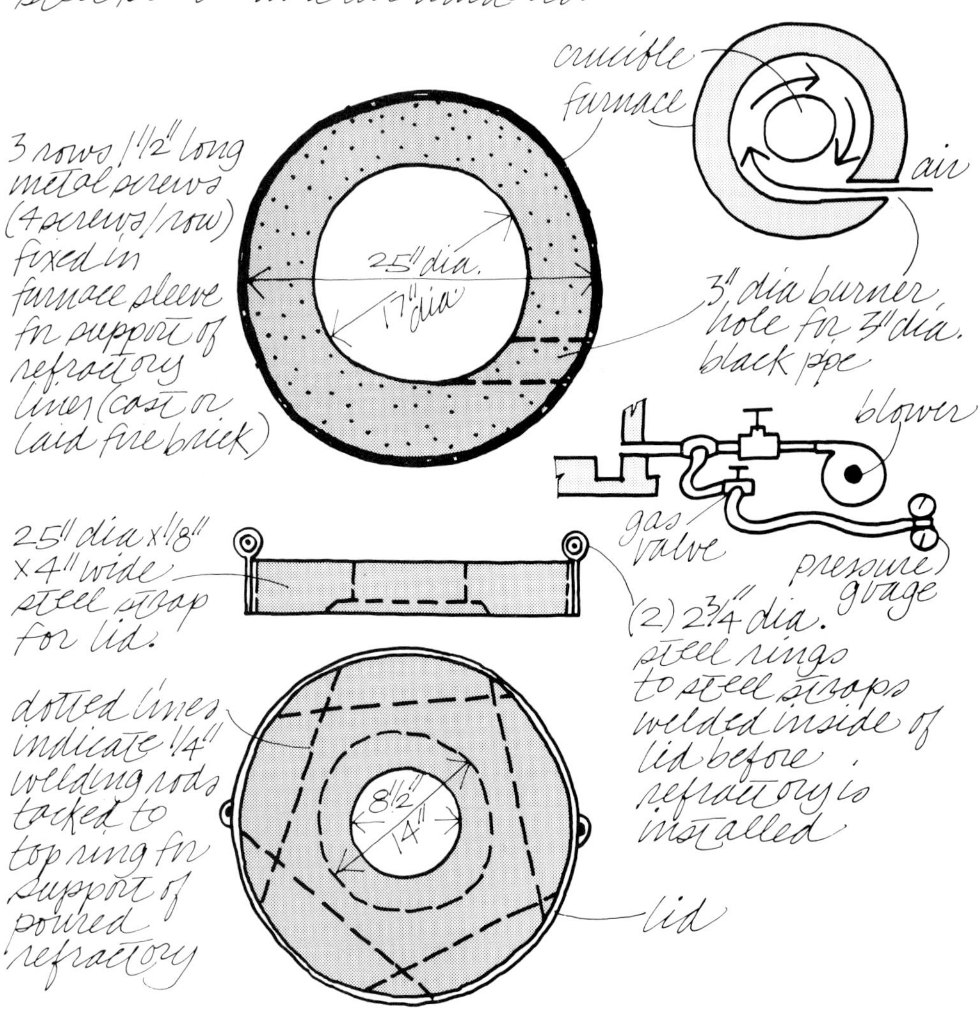

◄

Plans for One Furnace Design

California sculptor Robert Thomas shares his plans for an economical bronze melting furnace that can be built for under $150. Its capacity is a number 30 standard crucible (90 pounds of bronze). It is very durable and performs remarkably well.

The materials required are: one steel sleeve, #16 gauge galvanized steel 25 inches in diameter by 21 inches high; 1 sleeve 17 inches in diameter by 21 inches high; 1 galvanized steel ring #16 gauge 27 inches in diameter by 4 inches high (this is the bottom ring of the furnace); 1 ring of ⅛-inch-thick steel 25 inches in diameter by 4 inches high (this is the lid ring); 5 cubic feet of castable refractory—either refractory or firebrick are suitable and there are many others on the market—almost any castable that will take 2,400° to 2,600° temperatures would work.

The furnace sleeve, bottom ring, and top ring can be bent and welded by oneself or by a sheet metal shop and would cost approximately $20 to $40. The castable refractory would run about $75. In addition you need a blower motor and pipe for about $15 to $20.

For a blower use a 1,350 rpm blower and fix a butterfly air control in the 3-inch-diameter black pipe in front of it. Ahead of the shutter a hole is drilled in the 3-inch black pipe to accommodate a ¾-inch gas inlet pipe with a ¼-inch hole drilled in pipe cap to serve as the burner.

Crucible is being lifted from furnace by adjustable two-man crucible tongs, then lowered into the two-man crucible pouring shank. The metal is fluxed, skimmed, and poured. Ingot molds are visible. The crucible is a self-skimming type.

Investment Formula

Investment molds for lost wax are made according to the following materials and proportions.

plaster	2 parts
fireclay	1 part
silica flour	1 part
sand (30 mesh)	1 part
grog (8 mesh)	1 part

This is the formula for the original batch. Once you accumulate a certain amount of fired mold material it is not necessary to mix this batch again, since from this point you use *luto* (crushed used fired mold material). The mix used is then 9 parts luto, 3 parts plaster of paris, and 1 part every other time or so of firebrick grog just to keep the mix porous.

For the facing or first layer of investment on the wax use 3 parts silica flour and 2 parts plaster built to ¼ to ½ inch thick. Then burn out molds at around 700° F. and never go over 1,000°. These molds hold up very well at this temperature range.

Mr. Thomas uses only silica bronze (Herculoy) since after long practice he finds it offers far fewer problems for clean casts. It is easily welded with Everdur welding rod, and if you are casting large pieces in sections to be joined after casting, this is an important consideration.

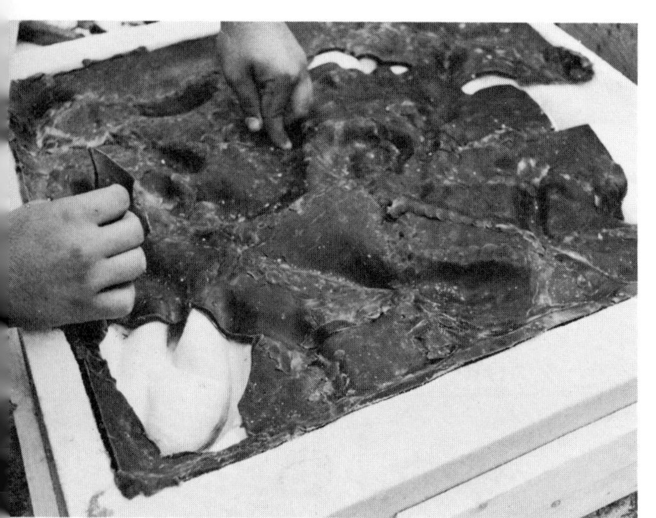

◄

A NO-BAKE SAND BINDING SYSTEM

A student presses wax into one-half of a sand mold. The original was made of plasteline clay, which was dusted with silica flour to prevent the sand from sticking. The clean silica sand of 50 to 60 GFN (Grain Fineness Number) was packed directly against the face of the relief. The binder was Foseco's Carset 500, which is basically sodium silicate mixed at approximately 4 percent the weight of the sand used. The catalyst or hardener is a mild ester used at 10 percent of the weight of the binder; both binder and hardener are safe and externally nontoxic. The sand, binder, and hardener are mixed together in a cement mixer with chunks of steel thrown in to assure good mixing.

After one side of the mold hardened it was turned over and the original removed. A quarter-inch layer of wax was pressed into the mold, the second half rammed over the wax and walls of the first half and, upon hardening, removed. The wax was then removed, sprues and vents cut into the sand, and the two halves glued and wired together with a mix of clear lacquer and iron oxide. The space left after the wax was removed will be filled by the metal. The mold breaks away easily from the casting upon cooling.

Photographed at Massachusetts College of Art, Boston

◄

McEnglevan model C 10 showing crucible furnace for aluminum or brass which is industrially made for schools, industry, or individuals.

STUDY FOR OMINOUS IKON. Dennis Kowal. 1970. Bronze, 17 inches high. A bronze study originally made in clay from which a latex flexible mold with a plaster support mold was made. A wax casting was then made from the latex mold and invested with ½ silica and ½ molding plaster, by volume. The spruing system was attached to the inside and the core vented; as a result no sprues or vents had to be cut from the surface of the piece, only from the bottom.

Courtesy, Jaffe-Friede Gallery, Hopkins Center, Dartmouth College

Appendix

NOTES ON COLOR PAGES

Notes for jacket front and color page opposite 56.

Eve. Tom McClure. Bronze and epoxy, 75 inches high.

Courtesy, artist

Tom McClure has developed a personal way of combining bronze casting and epoxy using traditional methods. He begins by modeling directly in clay, from which he makes a plaster mold. Wax castings are taken from the plaster and reworked. He also takes rubber and some plaster molds from objects such as typewriters, gears, wheels, and plastic toys, and then wax castings are made from these. In addition, molds are made of cylinders, spheres, and other abstract shapes resulting in a store of wax shapes that he combines with the figures. Castings are then combined directly in the final sculpture. For large pieces, some as high as 8 feet, braces are used to hold up the wax.

The finished wax is cut deliberately into segments for casting and according to the way an area relates to the finished piece; joints also are preplanned. The wax segments are cast by the lost-wax method or with bonded sand and then welded together. Sometimes, sheet bronze is integrated with the cast pieces and copper-nickel alloy castings with the ordinary bronze. Pieces may also be chrome- or nickel-plated. After the metal has been finished, epoxy is cast into the piece or pre-cast pieces are integrated with the total piece.

Breaker. David Black. 1967. Aluminum, 4½ feet high.

Photo, courtesy artist

To cast *Breaker* David Black has used the CO_2 sand process. His original was made from Styrofoam (stiff) or Ethafoam (flexible) sheeting. Cardboard templates and a hot wire were used to cut the foam into shapes. The foam original is packed directly into silica sand, mixed with saturated sodium silicate in a standard cement mixer, gassed with carbon dioxide (CO_2), and the aluminum poured hot (1900°) from the top down. No sprues are used on small pieces but on large pieces a big "well" sprue and bottom feeders are used which allow the metal to move quickly. The sand tends to chill the metal and the foam pattern helps hold the core in place so core pins are not needed. Sections are heliarced together.

Tribute to Those Magnificent Men and Their Flying Machines. J. Fred Woell. Epoxy, 29 inches high, 31 inches wide, 3 inches deep.

Courtesy, artist

Untitled. Roger Kotoske. 1971. Polyester resin; red, orange and yellow, 1 foot high, 3 feet wide. (See silicone mold series, chapter 2.)

Courtesy, artist

Notes for jacket back and color page opposite 57.

Forked Hybrid. Richard Hunt. Bronze, self-setting sand mold from clay. Edition of three, 58½ inches high, 44¾ inches wide, 24 inches deep.

Dorsky Gallery, New York

Untitled. Morris Applebaum. 1970. Colored epoxy and flexible vinyl. Vinyl was glued to epoxy casting.

Photo, Dennis Kowal

Ashford Medical Building. Puerto Rico. Rolando Lopez Dirube. Cast concrete. (See chapter 11.)

Courtesy, artist

American Woman. David Hostetler. 1968. Polished bronze, 41 inches high.

Courtesy, artist

Notes for color page opposite

Untitled. David Hendricks. Epoxy and leather. Epoxy was cast. The leather was pressed into the mold during casting.

Photo, Dennis Kowal

Untitled. Steve Daly. 1971. Cast bronze, aluminum, and iron, 5 feet.

Courtesy, artist

Swan and Its Wake I. Jack Zajac. 1968. Bronze cast in sand, 29¾ inches long. (See beginning of sand casting series, chapter 9.)

Courtesy, artist

Notes on color page opposite 89.

Students at the Hinckley-Haystack School of Arts and Crafts, Hinckley, Maine, pour aluminum into sand molds. Foam was used as the original pattern and is consumed and replaced by the entering molten metal. (See series on building a foundry, chapter 12.)

Photo, Judy Daley

Life-Size Plaster Waste Mold taken from original in clay. David Hendricks. Various colored epoxy sections will be cast into the mold and leather pressed in others. Epoxy is then cast over the leather and the other sections, and reinforced with fiberglass cloth.

Photo, Dennis Kowal

Eve. Tom McClure. Wax figure in progress. See text from *Eve* color photo on Frontispiece.

Bishop Panel. Tom Tasch. Epoxy.

Courtesy, artist

Tom Tasch casts epoxy in a plaster or rubber mold and reinforces the pieces with fiberglass. For color, he uses commercial pigments and epoxy from Thermoset Plastics, Inc. Sawdust is used for a filler when he wants some added impurities in a transparent color. Bronze powder was used for the helmet and the bishop's mitre. For mold release he uses Thermoset's paste wax and their PVA solution. The plaster mold is sealed with shellac prior to applying the wax.

IRON POUR CHART

TIME BETWEEN TAPS	CHARGE	TIME
	1-3	3:12
	4	3:18
	5	3:24
	6	3:27
	7	3:34
	8	3:44
	9	3:48
	10	3:56
	11	4:07
	TAP	4:10
	12	4:20
	13	4:25
45 MIN.	14	4:36
	15	4:40
	16	4:48
	TAP	4:55
	17	5:00
	18	5:09
	19	5:17
47 MIN.	20	5:25
	21	5:32
	22	5:37
	TAP	5:42
	23	5:46
	24	5:52
34 MIN.	25	6:01
	26	6:09
	TAP	6:16
	27	6:20
	28	6:26
34 MIN.	29	6:34
	30	6:44
	TAP	6:50
	31	6:53
	32	7:14
	TAP	7:22

CHART—Courtesy, Steve Daly (Professor) & P. McMahill (Asst.), Humboldt State College Sculptor's Foundry

MODELING CLAY

TO MAKE PLASTICINE (*non-hardening*) MODELING CLAY:

20 pounds microcrystalline wax (2300 grade Mobil)
1¼ gallons of Mobil #10 weight oil
7 pounds grease
50 pounds Gordon clay

Heat and blend wax, oil, and grease; mix in clay slowly and spread on wet plaster bat upon cooling. Mixture is ready for use.

Fine sand, etc., may be added or ingredients varied according to individual needs or preference.

Gordon clay or silica flour will make clay more plastic.

APPROXIMATE WAX TO METAL RATIOS BY WEIGHT

Wax to Metal

1 pound	—	9 pounds iron
1 pound	—	10 pounds bronze
1 pound	—	30 pounds aluminum

Density of Common Metals

Aluminum	2.702
Copper	8.92
Lead	11.34
Iron	7.9

APPROXIMATE WEIGHTS OF MATERIALS IN POUNDS PER CUBIC FOOT

Material	Weight	Material	Weight
Aluminum, cast	160	Nickel, cast	516
Aluminum bronze	475	Nickel-Silver	516
Babbitt metal	454	Oak (wood)	535
Bronze	534	Phosphor bronze, cast	536
Cast iron (mean)	450	Pine (wood)	30
Charcoal (lump)	18	Sand (molding), fairly dry and unrammed	75–85
Clay	120–135	Sand (silica)	85–90
Coal	60–80	Silver, cast pure	656
Concrete (set cement)	140	Steel (mean)	490
Copper, cast	548	Tin, pure	453
Fireclay	90	Water (as ice)	58.7
Gold, cast pure	1,200	Water (at 32° F.)	62.4
Gold, 22 carat	1,090	Wrought iron	480
Lead	710	Zinc, cast	428
Limestone	158–168		
Mercury	847		

WEIGHT OF CASTING FROM WEIGHT OF PATTERN

Allowance must be made for the weight of any metal in the pattern. The patterns are without cores.

A PATTERN WEIGHING 1 POUND WHEN MADE OF:	Will weigh when cast in: CAST IRON	CAST STEEL	YELLOW BRASS	GUN-METAL OR BRONZE	BELL BRONZE	ZINC	COPPER	ALUMINUM
Birch	10.6	11.2	11.9	12.3	12.9	10.2	12.0	3.9
Cedar	12.5	14.5	14.2	14.7	15.3	12.0	15.7	4.5
Cherry	10.7	12.0	12.0	12.6	13.5	10.4	12.8	3.9
Mahogany	8.5	9.5	9.5	10.0	10.5	8.2	10.1	3.1
Maple	9.2	10.3	10.3	10.6	10.9	8.9	11.0	3.2
Oak	9.4	10.4	10.5	10.8	11.0	9.1	11.2	3.4
Pine	14.7	16.3	16.5	16.6	17.3	14.3	17.5	5.3
Brass	0.84	0.98	0.95	0.99	1.0	0.81	1.04	0.31
Iron	0.97	1.09	1.09	1.13	1.18	0.93	1.17	0.35
Lead	0.64	0.75	0.72	0.74	0.78	0.61	0.8	0.23

MELTING POINTS AND SUGGESTED POURING TEMPERATURE OF METALS

METAL	MELTING POINT, ° F.	*Pouring Temperature Per Cross Section* UNDER ½ INCH	½ TO 1½ INCHES	1½ INCHES AND OVER
Aluminum	1218	1350	1310	1275
Aluminum silicon alloy		1350	1310	1275
Aluminum zinc copper alloy		1325	1300	1250
Aluminum magnesium alloy		1325	1300	1250
Aluminum zinc magnesium alloy		1325	1300	1250
Antimony	1166		1320 avg	
Babbitt, lead base	462		625 avg	
Babbitt, tin base	464		916 avg	
Brass, Muntz metal	1630	1890	1850 avg	1820
Brass, red	1952	2280	2190	2100
Brass, yellow 60–40	1858	2010	1920	1870
Brass, yellow 80–20	1877	2125	2050	1980
Bronze, aluminum	1922	2280	2190	2100
Bronze, tin		2280	2190	2100
Bronze, manganese		2280	2190	2100
Bronze, phosphor		2010	1970	1925
Bronze, silicon		2190	2120	2050
Copper	1982	2280	2200	2125
German silver	1850		2100 avg	
Gold	1945		2150 avg	
Grey iron, Class 20	2150		2550	
Grey iron, Class 35	2270		2625	
Grey iron, Class 60	2360		2650	
Malleable iron			2800 avg	
Lead	621		720 avg	
Linotype	486		620 avg	
Magnesium	1204	1420	1380	1350
Magnesium aluminum alloy		1475	1400	1350
Monel	2415		2750 avg	
Silver	1762		1950 avg	
Tin	450		650 avg	
Zinc	786		900 avg	
Zinc aluminum copper alloy		1110	1060	1025

Glossary

Accelerators
Chemicals that speed up but never start a reaction. Also called *promoters.*

Acrylic
A synthetic resin prepared from acrylic acid or from a derivative of acrylic acid. *Lucite* and *Plexiglas* are trade names of acrylic products.

Bentonite
A colloidal clay derived from volcanic ash and used as a binder with synthetic sands, or added to ordinary natural sands for extra strength.

Binder
An artificially added bond, other than water, used with foundry sand, such as cereal, pitch, resin, oil, sulfite, etc.

Binder; Plastic or Resin
Thermosetting synthetic resin material used as a bonding agent for core sand.

Bott (bot)
A mass of clay on a rod used to stop the flow of metal from the taphole of the cupola.

Bronze
Copper base alloys, with tin as the major alloying element.

Burner
A device that mixes fuel with air intimately to provide good combustion when the mixture is burned.

Burnout
See Lost-Wax Process.

Calcining
The preparatory heat treatment given raw refractory material to remove volatile substances and moisture, as in an investment mold: to calcine investment or plaster.

Calcium Carbonate
A solid material in nature as calcite and converted for use in making lime and portland cement.

Catalyst
A substance that accelerates a chemical reaction but that undergoes no permanent chemical change itself. In plastics, however, the catalyst can be combined and change its structure.

Cement
Finely divided mineral substances that harden as a result of chemical reaction or crystallization.

Chase
To finish the surface of metals by tooling.

Cire Perdue
See Lost-Wax Process.

Cold Molding Compound (CMC)
A flexible mold material: polysulphide liquid.

Core
A preformed aggregate or collapsible metal insert placed in a mold to shape the interior of the casting, which cannot be shaped by the pattern.

Crucible
A ceramic receptacle with high thermal conductivity used in melting metals.

Cupola
A small furnace, usually lined with refractories, used for melting metals.

Cure
To harden core by heating; to harden. Also called "set."

Deoxidation
Removal of excess oxygen from molten metal by adding materials with a high affinity for oxygen.

Dross
Metal oxides and other scum on the surface of molten metal.

Elastomer
A rubberlike elastic material such as natural or synthetic rubber.

Epoxy
A thermosetting plastic.

Epoxy resin
Resin used for making patterns. A mixture of resin and hardener sets chemically at room temperature.

Exotherm
A chemical reaction giving off heat; the amount of heat given off in a reaction.

Fiberglass
A material made of woven, spun, chopped, or matted fibers of glass made by spinning melted glass into filaments. *Fiberglas* is a trade name.

Filler
An inert substance added to a material to reduce cost or improve physical hardness and strength.

Firebrick
Brick made from refractory clays used in lining furnaces.

Fireclay
Clay capable of enduring high heat without fusing, used for firebrick and crucibles.

Fire scale
A metal oxide surface scale resulting from a hot pour.

Flash
A thin section of excess material formed at the mold along the parting line.

Flux
A material that causes other compounds to contact or fuse at a lower temperature than their normal fusion temperature. Material also used as a cleansing agent to dissolve or float out oxides that form on the surface of a metal during heating.

Gate
Portion of the runner or sprue in a mold where molten metal enters the casting or mold cavity.

Gating System (Sprue System)
The complete assembly of sprues, runners, gates, and individual casting cavities in the mold.

Gel
A jellylike material formed by the coagulation of a colloidal liquid.

Gelatin
A glutinous material obtained from animal tissues by boiling and extracting the protein.

Green Sand
A naturally bonded sand or sand mixture tempered with water; used damp or wet.

Ingot
A mass of metal cast to a convenient size and shape for remelting or hot working.

Investment
A flowable mixture poured around patterns that conforms to their shape and sets hard to form the investment mold.

Lost-Wax Process (Cire Perdue)
A casting process using a wax or thermoplastic pattern which is invested in a refractory slurry. After the mold is dry, the pattern is melted or burned out of the mold cavity.

Lucite
See Acrylic.

Methyl Methacrylate
See Acrylic.

Mold
The form, made of sand, metal, or other investment materials, that contains the cavity into which molten metal is poured to produce a casting.

Neat Cement
Portland cement mixed with water only.

No-bake Binder
A synthetic liquid resin sand binder that hardens completely at room temperature, used in cold-setting processes.

Parting Agent (or Compound)
Material spread on mold halves or patterns to ease separation of pattern or casting from mold. Also separating agent.

Patina
The color of the metal surface caused by oxidation, application of acids, or impurities.

Plastisol
A colloidal dispersion of a synthetic resin in a plasticizer. The resin usually dissolves at elevated temperature resulting

in a homogeneous plastic mass upon cooling.

Polyester
A thermosetting resin with a syrupy consistency.

Polystyrene
A water white thermoplastic composed of bonded beads used in casting for making patterns.

Pouring
Transfer of molten metal from furnace to ladle, ladle into molds, or crucible into molds.

Promoter
The chemical agent that speeds the action between various resin ingredients.

PVA
Polyvinyl alcohol.

Pyrometer
An instrument used for measuring temperatures above the average range of liquid thermometers.

Ram
Process of packing sand in a mold.

Refractory
Nonmetallic, heat-resistant material used for furnace linings.

Resin
A solid or semisolid mixture used as a binder. It has no definite melting point and shows no tendency to crystallize.

Riddle
A screen device used to remove coarse particles from molding sand.

Silica
Silicon dioxide, the prime ingredient of sharp sand and acid refractories.

Skimming
Removing or holding back dirt or slag from the surface of molten metal before or during pouring.

Slurry
A thin watery mixture of a clay-like dispersion such as liquid mud, cement, or mortar.

Sprue
See Gating System.

Thermocouple
A thermoelectric couple used to measure temperature differences.

Thermoplastic
Material that softens when heated and hardens as it cools.

Thermosetting plastic
A substance that becomes a solid when heated.

Vent
A small opening or passage in a mold or core to facilitate escape of gases during pouring.

Bibliography

AMERICAN FOUNDRYMEN'S SOCIETY. *Metal Casting Dictionary*. Des Plaines, Ill.: 1968.

ANGIER, R. H. *Firearms Bluing and Browning*. (Museum of Historic Arms.) Maine Beach, Fla.: Stackpole, 1940.

BIRINGUCCIO, VANNOCCIO. *Pirotechnia*. Cambridge, Mass.: MIT Press, 1966.

BRADLEY, J. H. *Materials Handbook*. 10th ed. New York: McGraw-Hill, 1971.

CLARKE, CARL DAME. *Metal Casting of Sculpture*. Butler, Md.: Standard Arts Press, 1948.

COWLES, FRED T. *An Elementary Foundry Manual*. Danville, Ill.: McEnglevan Co., 1967.

Foseco Foundryman's Handbook. 7th ed. New York: Pergamon Press, 1970. Compiled by Foseco, Inc., Cleveland, Ohio.

LOCK, D. FISH. *Metal Coloring*. Teddington, England: Robert Draper, Ltd., 1962.

MAYER, RALPH. *The Artist's Handbook*. 3rd ed. New York: The Viking Press, 1970.

MEILACH, DONA, and SEIDEN, DON. *Direct Metal Sculpture*. New York: Crown Publishers, Inc., 1966.

MEILACH, DONA Z. *Creating with Plaster.* Chicago: Reilly and Lee, 1966.

NEWMAN, THELMA R. *Plastics as an Art Form.* Philadelphia: Chilton Book Co., 1969.

NORTON, F. H. *Elements of Ceramics, Enameling on Metal.* Cambridge, Mass.: Addison-Wesley Press, Inc., 1952.

RICH, JACK C. *The Materials and Methods of Sculpture.* England: Oxford University Press, 1963.

ROUKES, NICHOLAS. *Sculpture in Plastics.* New York: Watson-Guptill, 1968.

SAVAGE, GEORGE. *A Concise History of Bronzes.* New York: Frederick A. Praeger, 1969.

SIMPSON, BRUCE. *History of the Metal Casting Industry.* Illinois: American Foundrymen's Society, 1970.

STRUPPECK, JULES. *The Creation of Sculpture.* New York: Holt, Rinehart & Winston, 1952.

TEFT, ELDON. *National Sculpture Center Publications.* Eight publications of proceedings from sculpture conferences. National Sculpture Center, University of Kansas, Lawrence, Kansas, 1960–1970 (Mimeographed).

UNTRACHT, OPPI. *Metal Techniques for the Craftsman.* New York: Doubleday, 1968.

WALTON, C. F. *Grey Iron Castings Handbook.* Cleveland, Ohio: Grey Iron Foundry Society, Inc., 1958.

Magazines

Cement. LaFarge Aluminum Cement Co. Ltd., London. Quarterly.

Concrete Construction. Concrete Construction Pub., Inc., Elmhurst, Ill.

Modern Casting. Buyers Directory Issue and monthly. November 1970, vol. 58, no. 5. American Foundrymen's Society, Inc., Des Plaines, Ill.

Plastics, Design and Processing. Lake Publ. Corp., Libertyville, Ill.

Plastics Technology. Bill Brothers Publ. Corp., New York, N.Y.

Plastics World. 270 S. Paul St., Denver, Colo.

Other Sources

"Brass Welding Facts for Foundrymen." Cleveland Flux Co., 1026–40 Main Ave., N.W., Cleveland, Ohio.

BYRZOSTOSKI, JOHN. "Patinas," *Craft Horizons Magazine,* 1965, New York, N.Y.

Foseco Foundry Practice. Technical Sheets. Foseco, Inc., Cleveland, Ohio.

How to Heliarc Weld. Technical Sheets. Union Carbide Corp., Linde Division, 270 Park Ave., New York, N.Y. 10017

Sources for Supplies

THE following sources are listed for your convenience. Consult your telephone classified book for local sources and distributors. No endorsements or guarantees are implied by the authors.

Bricks, Blankets, Etc.

Babcock & Wilcox
Refractories Division
12288 Merry Street
Augusta, Ga. 30903

Elipse Fuel Engineering Company
Gas Burners
Rockford, Ill. 61100

Burners for Furnaces and Kilns

Maxon Premix Burner Company, Inc.
Muncie, Ind. 44302

Ceramic Shell

Avnet-Shaw
91 Commercial Street
Plainview, N.Y. 11803

Concrete Form Release and Waterproofing

E. A. Thompson Company, Inc.
1200 Gough Street
San Francisco, Calif. 94109

Core Wire

Erie Wire & Steel Division
Erie Iron & Supply Corporation
1211 Walnut—P.O. Box 1068
Erie, Pa. 16512

Crucibles, Refractories, and Abrasives

American Refractories and Crucible Corp.
New Haven, Conn. 06473

Electro-Ferro Corporation
661 Willet Road
Buffalo, N.Y. 14218

Epoxy, Etc.

Ren Plastics, Inc.
5654 South Cedar
Lansing, Mich. 48909

Devcon
Danvers, Mass. 01923

Dow Corning Corp.
Midland, Mich. 48640

Fluxes

Cleveland Flux Company
1026–40 Maine Avenue, N.W.
Cleveland, Ohio 44113

Foundry Equipment

Industrial Equipment
Minster, Ohio 45865

McEnglevan Heat Treating & Manufacturing Company
700 Griggs Street
Danville, Ill. 61832

Fuel Oil Burners

Burnham Corporation
Industrial Burner Division
Lancaster, Pa. 17604

Microcrystalline Wax

Bareco Division
6910 East 14th Street
P.O. Drawer K
Tulsa, Okla. 74115

Roger Reed & Company
Reading, Mass. 01867

Alexander Sanders and Co., Inc.
Route 301
Cold Spring, N.Y. 10516

Motors, Controls, Tools

W. W. Grainger, Inc.
100 Lincoln Street
Boston, Mass. 02135

Pigments

Naz-Dar Company
1087 North Branch Street
Chicago, Ill. 60622

Plastics and Supplies

Ram Chemicals
210 East Alondra Boulevard
Gardena, Calif. 90248

Industrial Plastics
United States Plastic Corp.
1550 Elida Road
Lima, Ohio 45805

Polishes

Competition Chemicals
Iowa Falls, Iowa 50126

Du Pont Polishing Compounds
Wilmington, Del.

Polyester Resin

American Handicrafts-
Tandy Leather Corporation
(Local outlets)

Pyrometers

Gulton West Instrument Div.
3860 North River Rd.
Schiller Park, Ill. 60176

L. M. Marshall Co.
Station B
Columbus, Ohio 43202

Leeds & Northrup
North Wales, Pa. 19454

North American Refractories Co.
Cleveland, Ohio 44114

Refractory Cement

Kaiser Refractories
Room 1084, Kaiser Center
Oakland, Calif. 94604

Norton Company
Troy, N.Y. 12181

Resins, Mold Materials, Epoxies, and Miscellaneous Needs

Almac Plastics
26400 Groesbeck Highway
P.O. Box 247 South Station
Warren, Mich. 48090

Lance Gypsum & Lime Products
4225 Ogden Avenue
Chicago, Ill. 60600

Perma-Flex Mold Company
1919 East Livingston Ave.
Columbus, Ohio 43209

Resin Coating Corporation
14940 N.W. 25th Circuit
Opalocka, Fla. 33054

Thermoset Plastics, Inc.
5101 East 65th St.
Indianapolis, Ind. 46220

Safety Equipment

Wilson Products Division
ESB Incorporated
P.O. Box 622
Reading, Pa. 19603

Personal Environment Systems, Inc.
P.O. Box 800
Glendale, Ga. 91209

Sands and Additives; Resins

American Colloid Company
5100 Suffield Court
Skokie, Ill. 60076

Ashland Chemical Company
Foundry Products Division
2191 West 110th Street
Cleveland, Ohio 44102

Borden Resins & Chemical Department
50 West Broad Street
Columbus, Ohio 43215

Du Pont Mineral Sands—"Zircon"
Information, Room 10526
Wilmington, Dela. 19898

Swift Chemicals Company
1211 West 22nd Street
Oak Brook, Ill. 60521

United-Erie, Inc.
Erie, Pa. 16512

Sand Blaster

A.L.C. Company
P.O. Box 506
Medina, Ohio 44256

Foundries Material Company
5 Preston Avenue
Coldwater, Mich. 49036

Sand Mullers

Almar Specialty Machines Inc.
663 5th Avenue
New York, N.Y. 10022

Sand Riddlers

S.A.T. Plastics Corporation
6009 South Route 31
Crystal Lake, Ill. 60014

Static Eliminators

Nuclear Products—3M
St. Paul, Minn. 55101

Silicone Mold Material

General Electric–Silicone Rubber
Waterford, N.Y. 12188

INDEX

Italic page numbers indicate illustrations.